A Particle of Clay

The Biography of Alec Skempton,
Civil Engineer

Judith Niechcial

Whittles Publishing

Typeset by
Whittles Publishing Services

Published by
Whittles Publishing,
Roseleigh House,
Latheronwheel,
Caithness, KW5 6DW,
Scotland, UK
www.whittlespublishing.com

ISBN 1-870325-84-2

Printed by Interprint Ltd., Malta

Contents

'To see the world in a grain of sand'
William Blake, *Auguries of Innocence*

Introduction

Engineering is indeed a noble sport, and the legacy of good engineers is a better physical world for those who follow them – Ralph Peck

Biographies are usually written about eminent people in the world of art, literature, politics, film, theatre, media, or philosophy. We are interested in the way their lives interact with their artistic and professional achievements. We are told, sometimes more than we need to, about their loves, betrayals, and travels. We learn the stages of their development in their field, and this enriches our understanding of their achievements. With very few exceptions, this spotlight is not usually turned upon those who are eminent in the world of science and engineering. Alec Skempton, or 'Skem', as he is called by his friends and colleagues, is one of the most eminent civil engineers of the 20th century. The knighthood that he was awarded in Millennium Year 2000 is a recognition of this.

On the continent of Europe, engineers are accorded much greater status and recognition than in the UK. This could be a result of the English historical tendency to value aristocratic birth and wealth, the attributes of the leisured English 'gentleman', over the 'tradesman' whose endeavours have influenced the shape of our country. The professions of law and medicine are accorded prestige and financial reward and attract the brightest graduates, while the word 'engineer' means, to most UK citizens, something more mundane.

'Civil engineering has a long history. Although practitioners did not begin to describe themselves as "civil engineer" until the 18th century, the origins of their work lie in the construction skills of the ancient world and in the works of military engineers.' (Brown)

In 1828, when the fledgling Institution of Civil Engineers was applying for a Royal Charter, Thomas Tredgold was asked to give a definition of civil engineering that could be included in the draft. He wrote that civil engineering is 'the art of directing the Great Sources of Power in Nature for the use and convenience of man…' As the profession developed, this definition was later extended but it is still apposite today.

In 1950 Skem gave a broadcast on the BBC Home Service that explains, in clear layman's language, what a civil engineer does.

The job of a civil engineer is to design and build bridges and dams, canals and irrigation schemes, roads, runways and harbours, and so on. In all these he draws on past experiences, of course. But a good deal of science enters into the problems as well, and there are, I think, three definite ways in which science contributes to the practice of civil engineering. The first of these is what is called the theory of structures. This deals with the strength of building materials and the ways in which the engineer can calculate the stress and deflections of bridges and other structures. The materials are either man-made, such as steel and reinforced concrete; or natural materials such as timber and masonry, which have been carefully selected for use. In either case their properties can be studied and can be specified, on any particular job, quite accurately. Moreover, the various beams, columns, arches and so on are built to the designs of the engineer, so that the whole structure can be successfully treated in a fairly exact mathematical manner.

The second way in which science has come to the help of the civil engineer is in the study of what is called hydraulics. This deals with the flow of water, whether in rivers or canals; wave pressures on harbour walls; the transport of sand in estuaries and on beaches; and, with the design of hydraulic machinery, pumps and turbines. The properties of water can, obviously, be studied very exactly and, where the environment is man-made, as in a turbine, the problem can be treated with great precision. Where the problems arise from natural conditions, as in a river or estuary, you can't be nearly so precise, and a complete mathematical treatment is difficult. For that reason, engineers often build large tidal or river models in which they reproduce natural conditions and predict the full-scale behaviour from observations made in the model. Hydraulics and the theory of structures are both long-established parts of civil engineering science.

Skem then goes on to describe his own branch of civil engineering, soil mechanics.

This deals with the physical properties of geological materials from the point of view of civil engineering, especially the clays, silts, sands, and peats. The practical problems tackled in soil mechanics include finding out what foundation load can safely be put on any particular ground; the design and construction of earth dams, of dock walls, and airfield runways; the study of landslips and, indeed, any job in which the natural ground plays an important part.

Now clays, sands and other types of soil are really rather complicated materials and their properties are considerably more difficult to understand scientifically than those of steel and concrete. Moreover, we have to take them as we find them. They occur beneath the surface, where they can only be observed by digging trial pits or by means of samples taken from boreholes. So it isn't to be expected that soil mechancis can be as precise as hydraulics or the study of structures.

Without this study and understanding of how soils behave when subjected to various pressures and conditions, buildings cannot be built on firm foundations, dams cannot hold water, embankments would collapse, railway cuttings would subside, and road and runway surfaces would crack. In other words, soil mechanics is necessary for the very fabric of our industrial society.

In his 1950 broadcast, Skem explains why it is that, in the Victorian age, structures and hydraulics became well-established sciences while, in soil mechanics, very little happened after the work of Alexandre Coulomb in 1846 on landslips in clay strata and earth dams, until the Austrian civil engineer, Karl Terzaghi, came onto the scene after the First World War.

A survey of the second half of the 19th century reveals the astonishing fact that there was hardly a single contribution of any lasting significance.

There are three factors that explain this, he argues. Firstly, the different economic consequences when things went wrong. Failures of structures, the Tay Bridge, for example, were extremely expensive to repair, needing skilled labour and expensive materials. On the other hand, if a dock wall collapsed, it needed only poorly-paid navvies to build it up again.

Secondly, as the traditional materials used in construction, timber and masonry, were replaced by cast iron and, later, steel, science had, of necessity, to be invoked to solve the new problems. On the other hand, the problems associated with foundations and retaining walls have altered more slowly. 'Geological materials are the same now as they were during the construction of the Roman bridges, the medieval cathedrals, and the early canals.'

The third factor is that 'the Victorians liked their sciences to be definite and exact. The properties of steel and water are more easily expressed in figures than those of earth and clay. It needed mental courage to place soil mechanics on a sound scientific basis, and this was Terzaghi's achievement.'

Professor Dick Chandler of Imperial College explains how the science of soil mechanics developed with the work of the Austrian civil engineer, Karl Terzaghi, who is acknowleged to be the father of modern soil mechanics. 'The 19th-century engineers were perplexed. They could excavate a slope in a clay and it would stand perfectly well, perhaps for several years, and then all of a sudden it would collapse. The basic reason for the delayed failure stems from Terzaghi's theory of consolidation and the principle of "effective stress". We now know the complete process. The understanding of the principles of effective stress and how it applies to foundations, slopes, piles, any area of consolidation is crucial to the understanding of what will happen when you build anything. If you are to make any sort of prediction, or do any rational design, you have to invoke the principle of effective stress, so it is the foundation stone on which everything else in soil mechanics has evolved.' (Personal communication)

The eminent French engineer, Jean Kerisel explains the ideas underlying Terzaghi's principle of 'effective stress'. 'Water is incompressible, whereas the conglomerates in which it stands may have very different compressibilities and permeabilities depending on their structures.' (Personal communication) Through his work and teaching, Skem, in particular, together with his team at Imperial College, has developed Terzaghi's theory, particularly in the understanding of the geotechnical properties of clay and the stability of slopes, and the application of this theoretical knowledge to the analysis of many real situations.

With the possible exception of the Mangla Dam in Pakistan, when people ask, 'What did your father do?', I cannot point to a monumental construction, a Millennium Wheel, a Humber Bridge or a Canary Wharf, and say, 'That is his achievement.' Work towards the increase in knowledge of the properties of soil and its behaviour under different conditions is considerably less eye-catching. But the men who build the monumental constructions depend on Skem's fundamental research.

His achievements are celebrated by engineers. As his daughter I am interested in his scientific and historical work but also in him as a father, a significant influence in my childhood and my life. My hope is that others will be interested not only in what he contributed to engineering knowledge but also in what sort of a man he is, what influences have formed him, how his ideas have developed, to what extent he is a product of his time, and to what extent he has influenced his time.

In compiling material for this biography, I have pieced together memories and impressions from very many friends and colleagues, some of whom knew him at certain stages of his life, some of whom have known him for decades. Some had reservations about aspects of him as a man, most liked him very much, but all admired him and expressed an enormous respect for his achievements, particularly in the immediate post-war years, when soil mechanics was a fledgling academic subject. It was a time of huge energy and a sense that almost anything was possible, now that the dreadful war years were over. A time for new building, in the metaphorical as well as the physical sense. He built up the first University Department of Soil Mechanics in England at Imperial College, London University, and made major advances in the understanding of the properties and behaviour of clay and of the stability of slopes. He was a Fellow of the Royal Society at the age of 47 and is now the grand old man of British soil mechanics.

Skem is the most thorough and painstaking researcher. He follows all lines of enquiry, exploring the by-ways down which these enquiries lead until he is sure he has extracted the important points. He checks and double-checks the factual details. If something does not fit, he will never gloss over, or evade the issue, but will pursue it doggedly until the dissonance is resolved. Over most of his life these researches have involved soil mechanics. He has, over his long and productive career, written no less than 127 academic papers and nine books, some jointly written and some edited. The majority have been on soil mechanics and geotechnical subjects, but he has also written extensively on civil engineering history and the lives of engineers. At one stage in his life, he also researched and published on French baroque composers (more of these later). A lesser-known production is a typewritten volume, the culmination of years of work, assembled in an A4 ring-binder, entitled '*For Judith and Katherine*'. In setting out the story of my father's origins, I am much aided by this research into his and my mother's family histories.

He tackled this project with no less enthusiasm and application than he applied to his published writing. The outcome is a factual but also absorbing and well-written account, accompanied by a companion ring-bound volume of copies of the extensive collection of family portrait photographs, the oldest of which were taken in 1855, and snapshots from 1903. These he has assembled from various relatives. Not only has he identified the people in the photos, he has fully researched the techniques used in early photography and the different methods of production of, for example, ambrotypes, albumen prints, and *cartes de visite*. Skem consulted the head of the Photography Department of the Victoria and Albert Museum about the technical aspects of these photographs. He told me that this expert had been impressed by the variety of early photographic techniques represented in the collection and the fullness of the family records. Skem's ancestors were enthusiastic and skilled amateur photographers.

I shall start this book with some family background and childhood influences. Skem's life divides into distinct phases and chapter divisions reflect this.

Judith Niechcial

This biography was written while Alec Skempton was alive; indeed he read through the manuscript before it reached the publisher. Sadly, Skem died before the book could be published, but both the author and publisher felt it appropriate to retain the immediacy of the present tense in this account and in so doing help prolong his presence.

Acknowledgements

*T*his book could not have been written without the enormous encouragement provided to me by Professor John Burland of the Department of Civil and Environmental Engineering, Imperial College, London University. His help has been especially invaluable in relation to the technical aspects. To do full justice to his contribution he should really have been cited as co-author. His colleague, Professor Dick Chandler has also been particularly helpful, allowing me to tape-record a conversation he had with my father.

Joyce Brown, Skempton's former researcher/personal assistant and later member of staff in the Department, has contributed a poetical perspective, and I have drawn considerably on her book, *Civil Engineering at South Kensington*. Ann Barratt, the college archivist, and Suzannah Parry of the departmental library have kindly unearthed material for me. Dr Arthur Penman has been most generous in sharing his own research into the history of the Building Research Station (now the Building Research Establishment). At the Institution of Civil Engineers, Mike Chrimes, the librarian, has provided information, enthusiasm, and support. Gordon Eldridge, late of Binnie and Partners, has spared no effort in reviewing and revising my drafts of the Mangla and Kalabagh dam passages, and I am most grateful to him. He and another Binnie colleague, Roger Brown, have supplied me with reports and correspondence relating to the Pakistan projects.

The Norwegian Geotechnical Institute in Oslo holds valuable resources, in the form of the libraries of the great engineers Karl Terzaghi and Ralph Peck. Suzanne Lacasse, the Director, and Elmo DiBiagio of the Institute have provided me with generous access to this material, and warm hospitality. I was also kindly given access to correspondence between Skem and the Institute's founding director, Laurits Bjerrum, one of Skem's closest friends.

Suzanne Ewart of Salisbury Cathedral library has also carried out research for me.

I am also most grateful to the following for their time and care in recalling past times and looking out materials for me: the late Bill Allen, Professor Nick Ambraseys, Roger Brown, Professor Angus Buchanan, Marjorie Carter, Rosemary Chorley, John Clark, Julia Elton, Professor Bob Gibson, Richard Goodman, Rupert Hall, the late Professor Sir Alan Harris, Dr. David Henkel, Professor John Hutchinson, Professor Jean Kerisel, Carlotta Lemieux (née Hacker), the late Frank Newby, Professor Ralph Peck, Dr Don Roberts, James Sutherland, Malcolm Tucker, Carlo Viggiani and Jane Walbancke.

Non-civil-engineering friends who have kindly spoken or written to me of their relationships and memories of Skem are: Gudrun Bjerrum, the late Nona Crowe, the late Pamela Freeman, Sheila Glossop, Ursula Mommens (formerly Trevelyan), James (Jas) Pirie, Sonia Rolt, Janet Roscoe, and my sister Katherine Sisum.

Through his enthusiam for the project, my publisher, Keith Whittles, has given me the incentive to bring the book to completion. Paul and Susan McQuail and my son, Alexander Stevens, of the University of Kent at Canterbury have provided translations, editorial improvements, and wise counsel.

I am most grateful to all these people for their help and encouragement. Any mistakes that have been made are mine, not theirs.

Family tree

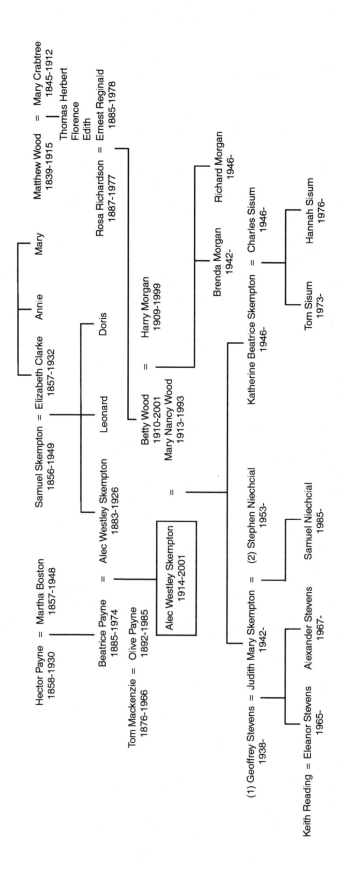

Chapter 1

Northampton

*H*e has a handsome bushy white moustache and a head of thick white hair, a stiff Edwardian collar and, although it is a warm summer evening, he wears a buttoned tweed suit and a tie as he closes the door behind him after Sunday tea to walk briskly to the Mount Pleasant Baptist Church on the Kettering Road for evening service. His name is Samuel Skempton. It is 1917. All the young men are away at war in the trenches of Flanders, including his dear son, Alec, of whom he will be thinking during the service. The house is *Hillside*, a distinctive and newly-built semi-detached white house with mullioned windows on the Wellingborough Road, Weston Favell, a village near Northampton. While Samuel crosses the fields and the park, his wife, Bessie, a handsome woman in the high-necked white blouse that was fashionable when she was young, clears up the tea things, and her daughter, Doris winds on the film in the Kodak pocket camera with which she has just

Mount Pleasant Baptist Church

taken a family photo. Bessie's daughter-in-law, Beatrice, Alec's wife, plays an energetic game of ball in the garden with her little boy, aged three. He has a shock of black hair and a shy smile. Bea, having been brought up in America, calls him Junior, as he is named Alec, like his father. Bea and Junior are staying at *Hillside* for a month this summer, because their narrow Victorian semi in Abingdon Avenue seems cold and lonely without Alec. Junior will grow up to be Professor A.W. Skempton and Samuel is his grandfather.

Skem's background is thoroughly English; the genetic pool from which he comes seems to have been completely unmixed with immigrant influences of any kind. Families stayed in the same group of small villages in the same part of the country in a way that was common in pre-Industrial Revolution England. They lived in Northamptonshire and Hertfordshire. The phrase 'Middle England' is particularly apt in describing this background, in a geographical and in a sociological sense. Skem's family is not aristocratic or wealthy but nor is it grittily working-class in origin. His forebears led hard-working serious lives, as small farmers, as rope-makers, or wheelwrights, or in the leather industry on his father's side and as maltsters or grocers on his mother's side.

This background is of central importance for Skem's identity. Frivolity, decadence, lassitude play no part in his make-up, nor do indigence or poverty. Diligence, application, hard work, and thrift play a central part. What is important is making a contribution, playing your part in the scheme of things, achieving the most you are capable of. These solid virtues stand behind him and his achievements. This is what it has meant for him to be a man.

Skem values his inheritance enormously, which justifies my laying it out in some detail.

Grandparents

Skem's Edwardian grandparents moved slightly up the social scale from their Victorian ancestors. A great-uncle donned a white collar as an accountant. The paternal grandfather, Samuel Skempton, who walked to Mount Pleasant that summer Sunday and many other Sundays, was a veritable pillar of Northampton society. He had been active in the building and setting-up of the Mount Pleasant Baptist Church and was deacon there for no less than 60 years. In his working life, he was Secretary of the Northampton Building Society. In his family history, Skem wrote:

> I remember him as a quiet courteous man, kindly but strict, saying grace before Sunday lunch but enjoying a good laugh, especially with my mother, of whom he was very fond.

Samuel's brother George was also a deacon at Mount Pleasant. On Samuel's retirement as treasurer in 1912, the *Northampton Echo* wrote of the brothers: *It is evident that the name of Skempton is one worthy of honour and one that will live long in the history of the church.* Of Samuel's career at the building society, Skem writes:

This job suited him admirably ... he had the integrity required in such a position of trust, and, as an institution conducted for the benefit of all within its ranks, not primarily for the making of profit but with strict adherence to sound finance, the Society measured up to his ideals.

Samuel eventually became a director, a post he held until 1947, exactly 50 years after he first joined the society. In 1912, he had moved up-market from the narrow terraced house in Holly Road to the much more spacious *Hillside*, where Skem later stayed as a small child with his mother, Beatrice, while his father was in the trenches. It was for Skem 'a real family home, of which I have fond memories'.

From as far back as he can remember, Bea and Junior used to walk from their Abington Avenue house near the county cricket ground at Northampton to *Hillside* for Sunday lunch and, in the summer, for tea in the garden. (Sixty years later, Skem follows that pattern of Sunday visits, but now to my house in Beckenham.) In the other direction, Samuel called into Skem's mother's house for coffee on his walk into town once or twice a week, a walk he continued into his 80s. Daily walking is another family characteristic Skem has inherited. He has walked the mile between his Kensington flat and Imperial College and back daily since 1947, and it is to this form of regular moderate exercise that he owes, I think, his good health and long life.

Samuel met his wife Elizabeth (Bessie) Clarke at the Baptist Chapel. Before her marriage, she had worked in the shoe trade. Northampton's main industry was leather working and several other forebears had been leather workers. She was particularly determined that her children should be well educated and, to this end, taught them herself when they were young. Skem remembers her as kindly but reserved. He writes,

I was usually on my best behaviour in her company ... she expected intelligent people to have good manners and respect for others; just as she expected and encouraged them to make the most of their talents.

Skem's own demeanour demonstrates these characteristics and I could not find a better way of summing up his expectations of his own grandchildren.

Skem is the most unmaterialistic person but his great sense of family means he values intensely the things of beauty he owns as a result of their passing through the generations. Skem inherited, through his Great-Aunt Annie, (Bessie's sister, and retired postmistress of Witney, Oxfordshire) a large and ancient carved-oak blanket chest, which had been handed down to the eldest of each generation. Skem's middle name, Westley, is the maiden name of Bessie's and Annie's mother, Mary.

Samuel and Bessie's eldest child, born in 1883, was Alec, Skem's father, after whom he was named. A second son, Leonard, followed two years later, and a daughter, Doris, was born in 1892. Uncle Len spent most of his life, apart from time serving in the 1914–18 war in France, working for the Royal

Insurance Company in Birmingham. Len and his wife, Gladys, had no children and, when he retired in 1948, they bought annuities for all their capital, which enabled them to live quietly for many years at the Crag Head Hotel, Bournemouth. Len did not have much contact with Beatrice and Skem. I remember there being a sense in the family that this couple were somehow not quite the thing, a bit snobbish, to be living in a hotel, playing bridge, and taking afternoon tea. It was a lifestyle with which Beatrice and Skem had little sympathy. When Skem later talked to his aunt Doris, she said that Len was in fact very proud of him, but Skem had never got that impression.

Aunt Doris, however, played a larger part in Skem's life. He remembers her as lively, with plenty to say. I have always assumed that the reason why she never married was that so many men of her generation were wiped out in the First World War. She lived the life of a middle-class spinster. Like her mother, she worked in the leather trade, in the office of shoe manufacturers, Stimpson Brothers, and ran the household at *Hillside*, caring for her parents through their old age, until her father died in 1949. She enjoyed photography and was present during the Sunday visits to *Hillside*. Beatrice was friendly with Doris, her sister-in-law, and family visits for Sunday lunch continued after Doris set up her own home after her father's death, in a bow-fronted semi-detached 1930s house in Meadway, Weston Favell. In later life, she was a large cheerful chain smoker. When she could no longer live independently, Skem arranged for her to move to a residential home in Abington Park Crescent. She was his connection with Northampton during his adult life and his sadness at her death in 1985 was compounded by the realization that that link was now broken.

On the wall of Beatrice's similar semi-detached suburban house in Watford, which Skem has kept on after her death, there are two exquisitely embroidered pictures in silk of a shepherd and shepherdess, in their original oval gilt frames, circa 1820. These are by Skem's great-great-grandmother on his mother's side, Sarah Pryor. Her daughter, Lucy Maria, was also a needlewoman, although less skilled. Like many young girls of her time, she made a sampler to demonstrate her talents with the needle. Hers was made in 1832 when she was 14. She also undertook a larger project, an embroidered map of England but, for some reason, she forget to include the Isle of Anglesey, which gives it a primitive air. The pig's profile that is the outline of Wales lacks an ear. These treasures are also on the wall at Orchard Drive.

Lucy Maria married Thomas Payne, elder son of an old Northamptonshire Quaker family. Their younger son, Hector, was Skem's maternal grandfather. Hector Payne and Samuel Skempton became substitute father figures for Skem during the absence of his own father, Alec, in the trenches of Northern France. Hector played a far more important role in Skem's upbringing than many a grandfather, and certainly greater than Skem himself later played with his own grandchildren. Hector, like his grandfather and father before him, had been educated at Ackworth, the Quaker boarding school in Yorkshire, although he later left the Friends, finding the Quaker ambience of that period too restrictive. Hector was apprenticed as an engineer and ran a small engineering firm in Balsall Heath with a partner who turned out to be a rogue. Consequently, after his marriage to Martha Boston in 1884, and the birth of their elder daughter, Beatrice, later to be Skem's mother, Hector decided to try his luck in America. In 1887, the young couple set sail for Bridgeton, New Jersey with their little girl, Beatrice, and, two years later, moved to Clinton, Massachusetts, where their younger daughter, Olive, Skem's aunt, was born in 1892, and where Hector was involved in the development of electric-welded-steel fabric at the Clinton Wire Cloth Co. The company set up a factory in Warrington and, in 1909, Hector returned to England and took charge of the engineering side of this operation. The welded steel came to be used in concrete reinforcement and, in 1911, Hector was offered a job as works manager with British Reinforced Concrete Co. at Old Trafford. The company produced large quantities of this material for road and floor slabs, and in addition, Skem writes:

> By 1923 at least a hundred structures, at home and abroad, had been built on the BRC system: offices, industrial and public buildings, water towers, grain silos, small bridges and so on.

Skem knew from an early age that concrete was his grandfather's business. At the age of six, he was taken on a visit to the works. Mr Butler from the works, later Managing Director of BRC, was to play a very significant part in his life. Reinforced concrete was later to be the subject of Skem's own MSc thesis.

Hector's sister, Lucy, Skem's Aunt Lucy, played an important role in bringing Skem's father and mother together, and was subsequently a significant influence during Skem's childhood. She was unmarried and continued to live at Hertford House at Olton, near Solihull, where she and Hector had been brought up, until the death of her mother. She then inherited the embroidered treasures and furniture and silver from the Paynes and set up home in Kingsley Road, Northampton, moving in 1922 to Cheltenham, where she lived with a companion, Miss Clark, until her death in 1932. With characteristic fascination and scholarship, Skem has catalogued the family silver, not only from Aunt Lucy, but also that inherited from his wife Nancy's side of the family, complete with drawings of the initials engraved on the handles of spoons and forks. Skem and his mother frequently used to visit Aunt Lucy in Cheltenham for a week or so at a time. She made a deep impression on him and, he says, had a large influence on his personality. It is through her that he gained his very strong sense of family continuity. With her Quaker background, Aunt Lucy was a Victorian lady in the true sense of the word, courteous and with an impeccable demeanour. She was strict and had high expectations of behaviour. Skem found himself behaving differently at her house from the way he did at home. She and her companion invariably called each other 'Miss Payne' and 'Miss Clark' with punctilious formality, despite the fact that they lived together for years. (Skem never knew Miss Clark's first name.) At the same time, she was warm, intelligent, and creative. In her youth, Lucy had been an accomplished amateur painter in oils. Skem remembers two rural landscapes of hers hanging in Hector's flat in Torquay, and several in Cheltenham. For some reason, these were sold in the early 1930s, when the contents of her house were dispersed after her death, a decision later bitterly regretted by Beatrice and Olive.

As well as the silver and embroidery, she left a legacy of about £3000 to Beatrice, which could not have been more timely. Alec had died, and it arrived just as Skem was going up as an undergraduate to Imperial College, London University, aged 18. This was 'a welcome contribution to our somewhat limited resources at that time'. This is an understatement, for Beatrice would have had considerable difficulty supporting Skem at university were it not for this legacy.

Martha Payne…'the sweetest nature of any person I have known'.

Hector's wife, Skem's maternal grandmother, Martha Payne, was known as Pattie by family and friends, and as Ninny Grandma by Skem as a child, and that's what I called her. She was born in 1857 in Olton, where her father had a prosperous coal-merchant business on the Grand Union Canal, and where she met Hector. Skem says, 'She had the sweetest nature of any person I have known.' I remember her as an old lady just as she is in her portrait, tiny, with a velvet choker round her neck, sitting quietly with hands neatly crossed on her bag and feet in their barred and buttoned shoes. Quite a contrast to her rounded and rumbustious daughter, Bea.

In 1922, Hector retired and he, Martha, and Olive moved to Torquay, to a large flat in a tall white stuccoed house in St Luke's Road overlooking Torbay. The flat had large light rooms and, although it was on the first floor, it was approached by a door at ground level because of the steep hill. Hector loved sailing and went down every day to the harbour to sail and look after his 20-foot boat. He also had a 14-foot clinker-built dinghy, which he allowed Skem to use as his own. Skem went to Torquay every summer while his father was still alive and, after his death in 1926, Skem and Beatrice spent all of every single school holiday there until Hector died in 1930. They were happy times and Skem remembers his grandfather with much affection. Martha (Ninny) then came to live with Beatrice and Skem in Northampton and thence to Watford, where Skem's wife, Nancy, painted her portrait in 1941. She died in 1948, aged 91.

Hector Payne in his boat.

Beatrice

Skem's mother Beatrice, known to all as Bea, was an extremely loving, vivacious, energetic and, to Skem and many of his friends, formidable woman. When she left Clinton High School, Massachusetts, aged 18, she wanted to become a nurse but found that she could not start the training until she was 21, so, as for so many young people of today, travel was the idea. She came to England in 1904 to stay with Aunt Lucy in Northampton. There she met the handsome Alec Skempton and some of his family. She and Lucy visited Paris, and then returned together to America, where she started her training at Massachusetts Homeopathic Hospital in Boston. Her beauty and outgoing warm personality comes over in photos of her in the wonderful, if restrictive, Edwardian clothes of the day, with her smiling eyes, her hair piled up in elaborate style, her corseted waist, and lace blouse. She played in the hockey team, and I remember her telling me the discomfort, before the era of the bra, of strapping her breasts into a corset before running round the hockey pitch. After graduating in 1909, she stayed on at the hospital and became a head nurse in the children's department. Keeping in touch with her new friends, the Skemptons, in England, she sent some photographs of herself to the family, including one in her nurse's uniform from 1909.

Meanwhile, Hector and Martha had come back to England and settled in Old Trafford. Olive, who by now had also left school (the Greenwich Academy

Beatrice as a nurse.

in Connecticut), joined them and, the same year, 1911, Bea married her Alec. The young couple took a house in Abington Avenue, Northampton and, except for the period during 1917 and 1918 when Alec was in the trenches, this was Skem's home until 1936. He writes:

> My mother had a wide circle of friends in Northampton, some of whom she met as a member of the ladies' hockey team which played in the grounds of Delapre Abbey, now the County Record Office. Chief among them was Isobel Briggs, mother and aunt respectively of my friends Tony Jackson Stops and Michael Robbins. Isobel lived at Boughton, just outside the town, in a lovely old house which we visited frequently. Also at Boughton was her niece, Christina Tebbutt, another of my friends. Christina and Tony were both so fond of my mother that they attended her funeral at Weston Favell in 1974... Close by, in Abington Avenue, were May Wooldridge... and Bertha Eggington (wife of the manager of the Liverpool and Globe Insurance Company) whose daughter, Nona, was and still is yet another of my friends. The list might be extended to a dozen names, for my mother was one of the most generous and vivacious people one could hope to meet.

I, too, was very fond indeed of Bea, my Grandma. She often looked after me while I was little and her love for me was unconditional, in a way that that of my parents, who had to be concerned about my development and progress in life, could not be.

Skem's Aunt Olive, meanwhile, served as a nurse in the Voluntary Aid Detachment (the VAD) during the First World War and, in 1923, moved with her parents to Torquay, where she became very close to Skem during the successive summer holidays. At the age of 34 she married Tom Mackenzie, a Scot by birth and older than her by some 16 years, who had retired to Torquay after a career as a sheep farmer in Australia. During the Second World War, Olive did sterling work with the WVS. She and Tom holidayed each year in the Highlands, at Tomintoul. She did not have Beatrice's good looks – her long face was not enhanced by hair that was combed rigorously into a bun at the back of her head. She was invariably dressed in tweed skirts and, when she went outdoors, in one of a succession of exceptionally severe masculine-looking felt hats, which often sported a decoration that nowadays would be seen as gruesome – a rabbit's paw, or a bunch of feathers from some dead bird. But she was a kind, lovely person. Skem and Nancy often visited her and Tom in Paignton.

> It was always a pleasure to stay with them and they had several friends, principally Rose Crosfield, whom we got to know well.

After Tom died in 1966, Beatrice and Olive usually had an annual week's holiday together somewhere like Harrogate or Cheltenham. I can imagine them energetically sightseeing in their tweed skirts and stopping off in tearooms for cakes and sisterly laughter. Their last joint outing was to Skem's daughter Katherine's wedding at Malmesbury in 1972.

Alec Skempton (senior)

About Skem's father, Alec, little is known within the family, as he died so long ago and Bea did not talk of him. Skem writes at length about his father's schooldays and sporting activity in Northampton, which sound amazingly similar to his own experiences. He may have consciously followed in his father's footsteps. Skem was 12 years old and still at prep school when his father died.

Skem's father was born in 1883 and went to Northampton Grammar School from the age of 13 to 16. The school was in its mid-Victorian building in Abington Square, with 100 pupils, paying fees of six guineas a year. He played cricket and soccer in Abington Park. After leaving school, he played cricket with the Mount Pleasant team and, having abandoned soccer, from 1902 was a regular playing member of the Northampton Rugby Football Club, then as now one of the best teams in the country. It is only on discovering this that I realized why Bea followed international rugby so avidly on television into her old age. After he left school, Alec worked in the leather trade, still living at home but, in 1903, he was posted to the newly opened London offices of Hamisch & Co., an American firm of leather merchants in Southwark, near London Bridge. He moved to digs in South London and went home to Northampton at weekends, spending his holidays sailing on the Norfolk Broads.

The next year, 1904, he met Beatrice when she arrived in Northampton from Massachusetts on the prolonged visit to her Aunt Lucy. He made an

Alec Skempton with Skem, 1915.

excellent impression on Bea. A photograph of him from that period, with his parents and sister, outside their small Victorian house in Holly Road, Northampton, shows a handsome young man in a three-piece suit, with fob watch and highly polished shoes. After Beatrice returned to America in 1906, Alec became manager of the Northampton branch of Hamisch & Co., a position that required an expert knowledge of leather and frequent train travel to shoe-making towns like Kettering. He must have represented quite a catch for Bea, with his respectable background and good looks. He and Bea were married in 1911. One reason he chose their house in Abington Avenue was that it was close to the County Cricket Ground. There they lived happily with the baby, Alec, who arrived on June 4th 1914, until the fateful day in May 1916 when he was called up for the army. Skem describes his childhood home:

> This house where I spent all my schooldays, was similar to a hundred others in the neighbourhood – including Holly Road. With a narrow frontage and a tiny front garden, it extended a long way back and had quite a good garden with a brick-built shed at the end and a garden door leading into an alley. On the ground floor were a front dining room, a 'middle room' used as a study, a kitchen, and a back sitting room; on the first floor my parents' bedroom, my bedroom, a bathroom, and a spare back bedroom; under the front and middle room was a cellar and over the corresponding bedrooms a large attic with a window commanding a view of the cricket ground. The attic was my playroom, later the site of a Bassett Lowke O-gauge railway and, from 1929, a laboratory-cum-study with workbench, shelves for chemical reagents, and glassware, a writing table, and bookcase.

Despite the modest size of the Skempton establishment, Bea employed a 'help', Mrs Stubley. Washday was a particular memory of Skem's. Mrs Stubley would arrive, fill the 'copper' with water, and light the fire underneath it. The copper was a container, which really was made of copper, about three feet across, with a curved bottom, fitted into a brickwork base. During the considerable time that it took to heat up the water, she would get on with other housework. The process of doing the laundry would take Mrs Stubley and Bea the whole day to accomplish, boiling and rinsing and mangling and hanging up the week's linen. (The BBC television series, *The 1900 House* has brought this process vividly to life.)

Father in the trenches

Alec left for France that October, 1916. He served in the Royal Garrison Artillery (as an NCO) and saw continued action from April to November 1917 at Vimy Ridge and Passchendaele. *The History of the Royal Regiment of Artillery* describes some of the horror of being an artilleryman at Passchendaele:

> The Germans rained a merciless fire day and night on the exposed British batteries below them. They drenched the gun areas with gas, using the

dreaded mustard for the first time... Many detachments were brought to near exhaustion by the strain of serving the guns hour after hour, day after day, night after night, while wearing gas masks... (They) had almost no proper rest or relief from 16th–31st July and were near total exhaustion when at the time when they should have been most alert... the weather was bad, observation poor, and most batteries were moving forward into the appalling conditions of no man's land. Casualties mounted as the German guns pounded the exposed British guns... Conditions at the guns were terrible, pouring rain, thick oozing mud capable of swallowing a man, a horse, or even a gun, made hopeless platforms. For shelter, a piece of corrugated iron over a water-filled shell hole was all that could be done. Ammunition had to be carried forward by hand over duckboards to the guns under constant shell and gas fire. The sheer courage and endurance shown by the exhausted, weary gunners at third Ypres was as great as had ever been shown in the Regiment's history, for the horror of it went on and on without respite until death itself was merciful.

A major who was there wrote, 'The mud is simply awful... the ground is churned up to a depth of ten feet and is the consistency of porridge... there must be hundreds of German dead buried here, and now their own shells are re-ploughing the area and turning them up. While we were shooting last night I saw a horrid sight. A Gunner from another battery ran through my guns, he was crazed with shock and, holding a hand blown from his wrist, he ran into the darkness and mud shrieking, never to be seen again.'

Alec somehow survived this inferno alive, but was gassed – however, not seriously enough to gain the respite of hospitalization. Skem takes up the story:

He then had a fortnight's leave before being engaged in the rearguard battles against the German advance, from March to August 1918, and in the vanguard of the Allied offensive from August to October... The trench warfare in 1918 amounted to eight months of unrelieved hell on earth. He suffered an attack of pleurisy in July during the retreat but had to recover as best he could in the dugout, and pleurisy returned immediately after the Armistice. This second attack he endured in billet, for even then there was no opportunity of going to hospital and, having recovered surprisingly well, as it seemed, he came home on leave in late December 1918, to be demobilised in February 1919.

Skem was one and a half when his father went to war, and only remembers him having these two weeks' leave during the three years he was at the front, so it is not surprising Alec is a shadowy figure to his son. The photograph (opposite), taken just after the war, of Junior looking proud in his father's tin helmet and medals is poignant in the light of later events.

Bea and the infant Alec (or Junior) spent very little time at Abington Avenue after 1916 until Alec returned from the war. They spent most of that period with her parents, Hector and Martha Payne, in Warrington and later Old

Skem with medals at Witney, 1918.

Trafford, with three visits of a month each to Samuel and Bessie at *Hillside*. Junior was a late developer as far as learning to talk was concerned. Bea told the story that, on their visits to Old Trafford, he was fascinated with the piledriving on the Manchester Ship Canal and, although he couldn't talk enough to ask to

At Hillside with his mother and grandparents, 1917.

go, the four-year-old Junior let her know that he wanted to be taken to see the piledrivers by imitating with his arms the up-down movements of the the machine. His interest in engineering was evidently *not* late in developing.

After the war, Alec returned to his job as manager of Hamisch and Co. In 1920, he took over the Northampton business, while retaining the name, and moved to new premises in Abingdon Street, where there were a couple of offices and a single-storey top-lit warehouse behind. He employed 'a secretary and a warehouseman, (who) also served as handyman and part-time gardener at the house'.

> Then followed three good years, made particularly successful with a very fine leather for slippers and ladies shoes from a tannery in Yeovil – a new venture. However, in 1924 his health began to deteriorate and in 1925 he became seriously ill with tuberculosis. He managed to do some work at home but by 1926 even this became impossible and he died in November. The doctor confirmed that the illness originated from the wartime pleurisy attacks, which in turn probably derived from the gas, but the War Office refused to admit liability, ironically on the ground that he had no hospital record, so my mother received no pension or compensation.

The injustice of this has left Skem with an abiding distrust and dislike of government and bureaucracy. I never heard Bea complain – she was not the complaining type – but Skem is angry on her behalf.

How important was the death of his father in Skem's life? Skem now says he remembers very little of him and says that his two grandfathers filled the gap. Returning servicemen in 1919 often found it impossible to describe what they had gone through to their wives, let alone to their children. Bea rarely spoke of him, perhaps because of the painful memories, and Skem says he did not ask her questions about him, sensing perhaps that she did not want to talk about him. I think he must have seen relatively little of Alec even after the homecoming from the war. Alec was either ill or in hospital, or otherwise working hard in his business. The gap in his life left by an effectively absent father has had its influence on Skem's own approach to his fathering of my sister and me. He experienced parenting predominantly from several strong women – Bea, Olive, the formidable Aunt Lucy, and his two grandmothers – for whom he was the focus of attention. Without a role model, the role of father must have seemed shadowy. Our father was always a distant figure for Katherine and me, involving himself in our academic progress but uncertain how to behave towards two small daughters.

As was the fate of so many of her generation, Beatrice was left a widow with a child to bring up and little chance of remarriage because so many of the men of her generation had been massacred in the 'war to end all wars'. She did not return to nursing – in those days mothers did not work. She cared deeply for her mother, Ninny, but I do not remember her being involved in the voluntary or committee work that occupied so many of her contemporaries. She must have had lots of surplus energy which she directed towards supporting and encouraging Skem. Any slacking in doing his homework, any tendency to hang around the house purposelessly would, I imagine, not be tolerated for an instant.

The family was singularly devoid of children of Skem's age. Olive was childless, as were Aunt Doris Skempton and Uncle Leonard. There were no siblings and no cousins. Junior was the focus of attention not only of his mother but of his grandparents on both sides and his aunts and uncle. I am sure no-one put overt pressure on him to succeed or implied that they would love and respect him any the less if he did not make much of his life, but the influence of all this loving encouraging attention must have been a considerable driving force for Skem.

During his spell in the leather firm and before he became ill, Alec had been able to save £3000 to £4000, a considerable sum in those days. With no war pension or earnings, Beatrice lived on these savings, no doubt prudently managed, for the rest of her life. They lived an economically straitened life, with no extravagances, most of their money going to educate Junior. The rent of the house in Abington Avenue was £52 per year.

Schooldays

At the age of six, Junior went to Wayneflete House Prep School, the fees of which Bea paid after Alec died. He remembers it as a very good school giving him a 'classical education'. Dr Charlesworth was the headmaster, an important figure around town. Then, in 1928, aged 14 Junior moved on to the same Northampton grammar school that his father had attended, now in a new building. Parents of children who had come to the grammar school from the council school paid nothing but his mother had to pay fees of £14 per year because he had been at a prep school. No account was taken of actual family income. They spent every school holiday with Bea's parents, Hector and Martha, in Torquay, at their expense, naturally, until 1930, when Hector died. After that, every summer was still spent in Torbay, but round the coast at Paignton, with Olive and Tom. This kind of family support was and still would be so important for a woman, however resourceful, bringing up a young boy on her own.

This need to be prudent about money has had a lasting effect on Skem. He has always lived frugally, staying with family on holiday, rarely going to restaurants or the theatre, buying few clothes. Even now that he is very comfortable, with a large flat in the best address in Kensington and considerable sums invested from his consultancy fees and my mother's estate, he has a need to maintain a deep financial cushion beneath him. The idea of making any inroads into capital is unthinkable. He can never quite believe that there will be enough for possible unspecified exigencies. He has absolutely no time for people who splash money about.

The children Junior played with were the children of relatives or of friends of Bea's. Nona Eggington was one of these. Beatrice and Nona's mother, Bertha, were young mothers together – both active, hockey-playing, intelligent, and determined women. Her family lived close to Abington Avenue. Nona was born four months after Junior, in the October of 1914. Like him, she was an only child, and they spent a great deal of time playing together. She remembers being bathed together with him as a child. They could have been brother and sister. As the only child of a widow, Junior was always the centre of attention. They had wonderful times in the Northampton attic, where he had his train set, later replaced by a drum kit. She felt he was very kind to let a mere girl of ten or 13 'fool about with' his trains or drums. She says he treated her as an equal. From her description of herself, I am not surprised. She was clearly an energetic and ambitious girl. By the age of 12, she had learned to ride a motorbike and her teenage ambition was to become a mechanical engineer, not a career of choice for many girls in the 1920s.

Nona and Junior used to watch cricket matches on the Northants grounds from the attic window. She still adores cricket. She used to go with him to Seaby's, a tiled dairy shop that sold the most wonderful ice cream. Nona's father, like Uncle Len, was in the insurance business. In 1924, when Nona was ten, his firm, and so also the Eggingtons, moved from Northampton to Bedford, but the two families maintained close contact. The Eggingtons used to 'chunter over' to Northampton in their funny little car, called a Swift. They used also to go to Devon on holiday and, one summer, Nona went over to Paignton and sailed out into the bay in the dinghy with Skem. Despite his warnings when they went about, she did the usual thing, got knocked on the head by the swinging boom, and fell overboard, feeling a fool. Apart from herself, Nona thinks Skem did not have girlfriends as a teenager. He was a 'male chap' with his rugger enthusiasms.

Nona remembers Bea as 'a loving person in every way', an excellent mother, fair and no-nonsense in the way she brought Junior up, and she would never let him flag in his school work. She was a strong person and just got on with life. Nona remembers her characteristically striding walk. All her considerable energy was spent on Junior, but not in an over-cherishing way. Nona remembers an idiosyncrasy of Bea's was that she always turned the paper sideways when she was writing. When Skem went to university, Bea wrote to him weekly, but the letters were so illegible that he had to save them up until he saw her at the weekend to interpret them.

Skem and Nona were also friendly with Tony Briggs (son of Isobel Briggs, another great friend of Beatrice's). He had been at prep school with Skem. Tony later changed his name to Jackson Stops, which was his wife's name, when he inherited from his father-in law the famous estate agency which still

Skem as a scout, pictured in the garden at Abington Avenue (above).

bears this name. His uncle, Richard Caple, was music critic of the *Daily Telegraph*. From him Tony, Skem and Nona got piles of review copies of wonderful records by Duke Ellington, Benny Goodman, Cab Calloway, and other American bands.

Skem retains a great fondness for Nona. She never became a mechanical engineer. She got a job as secretary to the Export Director of Vauxhall Motors and later in her father's insurance office, as it had been denuded of male staff by the war. She married Ralph Crowe, an architect and travelled with him to the West Indies and, when he was Professor of Architecture at Newcastle University from 1977–87, lived in a beautiful stone barn overlooking the valley near Hexham where she brought up her three children. Three summers ago Skem invited her to lunch at the Ark restaurant and, to her astonishment, the lunch was followed by the delivery of a huge bunch of flowers. She told him she thought he was brilliant. He denied this, admitting only to a capacity for concentration and enormous hard work. She doesn't remember him ever talking about feelings, not even about the death of his father but then, she allows, 'men of his generation were not demonstrative'. What he did have were wild enthusiasms, like his love of rugby.

The other successor to the train set in the attic at Abington Avenue, alongside the drum kit, was the chemistry lab. Skem set it up with a bench, a Bunsen burner, and a tap and basin, which drained into a bowl that had to be taken away to be emptied, and carried out all kinds of experiments. Two friends from grammar school, Dick Smethers and Ray Dickens, had also fitted out science labs in their attics. It must have been fashionable among adolescent boys before the age of television and computer games. According to Skem, Ray's lab, and presumably also his scientific ability, was 'inferior'. Beatrice recounted one disastrous occasion when Ray did a chemistry experiment in the bath, which went so wrong that the bath was stained permanently bright blue. Skem and Dick fancied themselves as physicists and took a delight in thinking they knew more about the subject than their teacher at the grammar school, who was, it seems, easily teased.

Another activity of his childhood and youth was the Scouts. Despite his dislike of the discomfort of camping, he progressed to become a King's Scout and proudly posed for a photo in the uniform in the garden at Abington Avenue.

Skem was bone-idle at school and, by his own admission, he was 'not scholarship material', but he was complacent in the knowledge that he was bright. He got good marks for history and loved physics. In most subjects he got good exam results without much effort. By the age of 14 he had decided he wanted to be a scientist but, even then, he did not work particularly hard. Of all his teachers at grammar school, the one he remembers most is the senior maths master. Maths is a subject in which, as he says, 'you cannot hide'. When he was about 16 this teacher gave him a thorough telling-off, saying he was a disgrace and he must mend his ways or suffer the consequences. The memory of the impact this made has remained with him and jolted him into working much harder. In some biographical notes which he compiled for the Royal Society he boasts, 'By the age of 16 I had worked through most of Fenton's *Quantative Analysis* (1926 edition)'.

In those days children did 'matric' at 16, and Higher School Certificate at 18. He did pure and applied maths, physics, and chemistry at Higher. He remembers sitting the exams at Nottingham, staying with his Uncle Len, but doing the 'practicals', which involved lab work, at the Imperial Institute in South Kensington.

Despite the loss of his father, Skem had a secure and happy childhood and adolescence. Bea and her family provided large doses of love, pride in his childish achievements, and unending support. He had a small circle of good friends, and healthy boyish interests – cricket, rugger, the Scouts, chemistry, and jazz. He had avoided that horror of the English upper classes, being sent away to an all boys' boarding school. His identity had been securely forged from the solid middle-England values exemplified by his family background.

It was his connection with grandfather Hector's firm, British Reinforced Concrete, which was to be significant in the next stage of his life. When Skem left grammar school at 18 it was 1932, the middle of the Great Depression. The prospects of employment for young men, however solid their educational achievements, were uncertain, to say the least. Bea would have known who to ask to be a mentor to her handsome, bright, but not yet hard-working son. So she got in touch again with Mr Butler at BRC. He remembered Hector Payne and told her that there would be a job for her son at the works after he graduated. Mr Butler had himself been to Imperial College and recommended it strongly. And so it was that Skem went as an undergraduate to Imperial College, University of London, where he was to spend most of the rest of his life.

Chapter 2

Student Days at Imperial College
1932–36

*T*he background to Skem's undergraduate years was the rise of Hitler and Churchill's warning to the country to rearm. Industrial unrest and unemployment dogged the workforce in Britain and, as Christopher Lee writes (Lee, 1999), 'It was not easy to sell the idea of rearmament to a nation that could not hear the sound of gunfire.'

For a grammar-school boy from a provincial town and from a small close-knit family, coming to London, the big city, could have felt daunting, and these events perhaps seemed distant. But Skem immediately created safe and tight domestic and geographical boundaries within which he and his friends moved in the London of the 1930s.

For his first two years at college, he lived in the red-brick hostel with Portland-stone facings that still exists in Prince Consort Road, next to Holy Trinity Church. Modern-day student successors to Skem go shouting to each other in and out of the courtyard, pushing their bicycles. Flyers are still stuck to the stonework advertising student drama productions and political meetings. The following year, he moved about a quarter of a mile to digs in Cromwell Place, not far from Gloucester Road underground station.

Later, during his postgraduate year, he lived in other lodgings in Prince's Gardens, behind the gardens of Brompton Oratory. (The college now owns three sides of Prince's Gardens, accommodating hundreds of students.) All his various landladies cooked meals for him and did all his laundry, so that he was able to move effortlessly from being physically taken care of by his mother to hostel life and then on to the housekeeping of these landladies. After a brief return to his mother, he then married. So he has never had to cook and clean for himself, in a way that seems perfectly natural to him and, perhaps, to most men of his generation and class. The world of Skem and his friends was geographically limited to an area between Kensington Gardens and the Thames at Chelsea. He sometimes ventured as far as Cork Street, to the art galleries, but otherwise lived his whole life in South Kensington. A visit to the slums of the East End, say, or to Islington, would have been a venture into completely unknown territory.

Imperial College had been created from the joining of three constituent colleges, the Royal School of Mines, City and Guilds of London Institute, and the Royal College of Science. The Department of Civil Engineering was 'sectionalised into specialities: Structures, Hydraulics, Highway Engineering and Surveying' (Brown, 1985). As an undergraduate, he studied in the Victorian laboratories on the top floor of what was then called the Huxley Building, now the Henry Cole wing of the Victoria and Albert Museum.

Skem sailed through his first year at Imperial College, finding it 'a doddle'. He had his Highers under his belt – the equivalent of the modern A level – while several of his contemporaries had arrived with only Matric, the lower-level qualification. He was therefore excused half of the first year of the civil engineering course. He did a course in metallurgy instead, at which he did rather well, and spent the rest of that year enjoying being a student in London.

A significant figure in Skem's student career at IC was his teacher, mentor, and Head of Department, Professor A.J. Sutton Pippard, who was still at Imperial when Skem returned to join the staff several years later. Pippard had held a chair in civil engineering at Bristol University from 1922–28 and had been involved in the development of airships with Barnes Wallis and in the investigation of the crashes of the R38 and the R101. However, Skem wrote in his 1970 memoir on Pippard for the Royal Society, that he felt that he was 'never devoted to aircraft as such, but rather to the structural problems they presented and the intellectual challenge of designing complex frameworks with minimum weight, but adequate factors of safety'. Pippard had come to IC in 1933. His book, *The Analysis of Engineering Structures*, was published in 1936. Skem found stimulation in his teaching. He writes,

> The lectures in structures and hydraulics were probably the most advanced and the most demanding of any being given in the country.'

Joyce Brown, in her history of civil engineering at South Kensington describes how 'his wit in the class-room and ability as a teacher were admired by students' and quotes another former student, who recalled 'the attention Pippard gave to his lectures which were carefully prepared and presented, his blackboards a model of neatness and clarity. On one occasion, after apologising to a class for the fact that he had had a cold the week before which might have affected his performance, he insisted on giving the whole lecture again.'

Apart from Pippard there were others on the IC staff at the time who were important figures in Skem's development. Professor C.M. White taught hydraulics, and it was his approach to research that led Skem to consider a career in research for himself. S.R. Sparkes gave the structures lectures, and Miss Letitia Chitty was a brilliant mathematician. Above all the geology teaching of Dr. F.G. (Frank) Blyth was an inspiration for Skem.

Describing the situation during his postgraduate year, 1936, Skem writes,

> Pippard and White almost lived in the laboratories, directing and carrying

Goldsmith's extension, Imperial College (painting by R.T.Cowern, 1962). From Joyce Brown's A Hundred Years of Civil Enginering at South Kensington.

out research work and inspiring a love of scientific enquiry in the minds of those students willing and able to respond.

One of those minds was Skem's, and Pippard's influence on his interests and his way of working was considerable. Apart from the engineering and research influence, my guess is that Skem learned some of his great skill as a lecturer from Pippard.

During Skem's first summer vacation, he went back to his mother's in Northampton and worked in an architect's office, helping with drawings, going round sites, and becoming interested for the first time in the history of architecture. It was during his second year that there seems to have been some sort of turning point for him. He found that he hated not knowing things. It seemed to him that he must find out about everything, go back to basics and learn all there was to know. He became fascinated by geology, finding inspiration in the lectures of Frank Blyth. Suddenly he began to apply himself and it was during this year that he established his lifelong pattern of daily work. It was not only into his university subjects that he suddenly wanted to enquire. He began to read widely, particularly ancient history and philosophy.

Under Pippard, surveying was a subject occupying four hours a week in the third year. 'An engineer's training is not complete without knowledge of how to use surveying instruments and of the more advanced techniques available. Indeed, in the first years of an engineer's working life the need for surveying expertise may well predominate over other special knowledge.' (Brown, 1985) Students carried out surveying practice in Hyde Park and, at an annual surveying camp, the different country locations. A photograph of 1934 shows Skem in the back row of a group of earnest young men in a field on a 'Civils Survey Camp'.

During his second summer vacation, again back in Northampton, with exams looming, he worked at his university subjects all day and every day. He took his work with him to swot all through the annual holiday at Paignton, where, from a snapshot of him with friends and a primitive version of a surfboard, he seems to have had at least some jolly times. He continued to play drums, not very seriously, during his first year at college, and remained interested in jazz until the time when he discovered classical music. Nona remembers visiting him later in 1938, when he was living at Oakley Street, and going with him to a Cab Calloway concert. (Cab Calloway, known as the Hi-di-ho Man, had a big band, which followed Duke Ellington's into the Cotton Club in 1931.) I don't know whether Skem played cricket at college, but he certainly did some boxing, which was a new venture for him. He represented City & Guilds in an Inter-Collegiate Boxing Competition as a heavyweight in March 1933. His main sporting interest, though, was still rugby, which he played out at the sports ground at Sudbury Town, Wembley, captaining the team in 1936.

A student friend, Jas Pirie, remembers the young Skem as a man's man, not having many girlfriends, being too busy concentrating on his studies. He was able to direct his attention very intensely on one aspect of work at a time, which Jas describes as his tunnel vision, excluding what he saw as trivial or marginal considerations.

The South Kensington Museum area is made up of that strange and wonderful conglomeration of cultural institutions created on land bought from the proceeds of the 1851 Exhibition. In the 30s, the Science Museum was

With friends on the beach.

Captain of the Imperial College rugby team, 1936.

under construction, just south of Imperial College, in Exhibition Road. Near to IC, in Prince Consort Road, is the Royal College of Music and, opposite that, are the steps leading up to the Royal Albert Hall. A few hundred yards away, then occupying buildings under the arch leading from Exhibition Road into the Victoria and Albert Museum, was the Royal College of Art, which was to play an important part in Skem's student life. Nearly all the engineering students were, of course, men and, where students at other male-dominated colleges looked to nursing or acting students for feminine friendship, the engineering students at IC had female art students within a few yards of their building. The RCA junior common room was the venue for Friday evening dances. He met his men friends at the hostel in Prince Consort Road and met women at these Friday 'hops'. It must have been a heady mix, the men with their scientific and technical education and sporting interests meeting with creative and artistic women, who, in those days, would have had to overcome considerable obstacles to live independently and study art full time at a prestigious college in London.

David Green was a member of the group of friends who were to be so important to him. David was a metallurgist studying at the School of Mines, who lived in the same hostel as Skem. He had been engaged to the stunningly beautiful Psyche, who was studying art at Chelsea College and David was devastated by her rejection of him. He later went on to marry Betty Bishop, who was not in the RCA circle – she was a high-flying secretary and was tough and assertive, while David was a gentle soul. Psyche married Jas Pirie, who was reading chemistry at the College of Science and was a pivotal figure in the circle. Jas reports that Skem was strongly attracted to Carol Dugan who lived in a mews flat behind Imperial College, but he thinks he never took her out. There was also Tony Smythe, a keen mountaineer, and member of the Alpine Club. There were others in the group who did not survive the war, among them Archy Roland, with whom Skem had shared digs in his third year in Cromwell Place.

One of the art students at RCA was Eleanor Haley. Skem was fond of her – she was 'a lovely girl' and the daughter of an upper-class family. Her father was some kind of senior diplomat, perhaps even the Ambassador in Ceylon. Although all the members of this student group were politically left-wing, Skem was surprised when Eleanor became a communist during the war, seeing it as uncharacteristic of her to go to such extremes. During the war, afraid that her politics would lead to unpleasant consequences, she hid herself away in Preston and, somewhat surprisingly, she married a milkman.

Eleanor shared a mews flat off Queens Gate with another student at RCA, a beautiful down-to-earth, brown-haired young woman from Yorkshire called Mary Nancy Wood (known to all as Nancy.) Nancy was very fond of Jas Pirie and painted a portrait of him. He paid a weekend visit to her parents' house in Yorkshire but this was not a success, as Nancy developed laryngitis. He then made the mistake of introducing her to the devastatingly handsome but serious and remote Skem. She knew immediately that this was the man she wanted to be her husband, and set out with some determination to win him.

…the devastatingly handsome but serious and remote Skem

Nancy's self-portrait. Photograph by Colin Crisford.

Nancy looks out of her self-portrait with clear brown eyes. She was the artistic younger daughter of Reginald Wood, a short stocky solid Yorkshire silk-mill owner, and his taller, rather morose wife, Rosa. Nancy had spent her girlhood in Brighouse, near Huddersfield, where her grandfather, Matthew Wood, had built up the silk-spinning business. When he took it over in 1890, the Thornhill Briggs silk mill had been a huge three-storey structure, fifteen bays in length with an office block and workshops. In 1910, Matthew more than doubled the size of the mill, adding a 20-bay extension and powering his machinery by a massive steam-mill engine. He was the epitome of the northern Edwardian entrepreneur. He brought in his elder son, Reginald's brother, Thomas Herbert, as a partner, who ran the business, by now called Wood Bros. & Sons, with brilliant success. Skem writes of Thomas Herbert,

> He had mastered every detail of the technical and financial operations and increased his own capital by careful, well-informed investments.

After Reg left school in June 1901, aged 16, he joined his father and elder brother at the mill.

> It was typical of Thomas Herbert that in December he set Reginald, after six months of training, a technical exam paper on the processes involved in silk spinning and required detailed answers in writing to all the questions,

Matthew Wood – the Northern Edwardian entrepreneur

writes Skem in the family history. Nancy's mother Rosa always said that Reg had been 'very good with the men'. He evidently excelled in what would now be called 'industrial relations' with the workforce, as his strict Yorkshire upbringing was tempered by a generous and kindly temperament. Reg married Rosa in 1909, and Nancy and her older sister, Betty, were brought up in their house, *Woodleigh*, which was 'one of the admirable Victorian houses in Huddersfield Road... stone-built, with high ceilings, heavy mahogany doors, marble fireplaces, and good-sized gardens front and back'. Thomas Herbert's business acumen led him to foresee 'the inevitable domination of artificial silk over the natural material and in 1922 (he) decided that the mill should be sold while still an active going concern. The sale was completed in June 1923, most of the proceeds being invested in Courtaulds'. Meanwhile, Matthew Wood had died in 1916, leaving an estate of £72,315 (more than three million pounds today). Most of this fortune was bequeathed equally to Thomas Herbert and Reginald, who thus became comparatively rich men. Thomas Herbert's son, another Herbert, had no interest in going into business with Reginald and, as Reg had no desire to work for anyone else, he decided to retire early too. He, Rosa, Betty, and Nancy moved to the Yorkshire seaside resort of Scarborough, where, in 1924–25, Reg built a second *Woodleigh*, another substantial, but a more modern, house. Rosa ran *Woodleigh* with the help of a cook-housekeeper and a daily maid, while Reg played golf on the course opposite the house. I remember him donning tweed plus-fours, a kind of trouser that ended in a band just below the knee, and long checked woollen socks. His Yorkshire work ethic did not allow him to live a life of complete leisure, however, and he served for many years as church warden of Scarborough Parish Church, St Martin's, South Cliff. Reg's life's work was the creation of a magical and beautiful hillside garden 'from what had been an acre of scrub and rough grass on a boulder clay slope', writes Skem, the geologist. In front of the house was a formal rose garden and, behind the house, stone steps and grassy paths led to a goldfish pond, herbaceous beds, vegetable plots, and soft fruit cages. A wooden gate at the top of the garden led into the steep woodlands of Oliver's Mount.

In Brighouse, Nancy had mostly been educated at home by a governess. In Scarborough, she and Betty learned to ride at Shaw's stable, where Reg kept his own horse in livery. She went to Scarborough Girls' High School and then to Scarborough Art School for three years.

Woodleigh, Scarborough.

When Nancy first introduced Skem to her parents, they looked upon him with great suspicion. He was, after all, a decadent southerner with a low salary and few prospects. (Betty's beau, the local dentist, Harry Morgan was regarded with greater favour.) Although Rosa was a rather self-absorbed woman, Nancy had a prosperous and secure childhood with a loving father who was present for much of the time. The legacy of this background was to make her a well-balanced, clear-eyed, sensible, and rather serious young woman. However, she stood apart from her forebears in her creativity. Skem writes in the Biographical Notes to his collection of her wood engravings (Skempton, 1989), 'She took the entrance test at the Royal College of Art in September 1933. A week later came the interview with Sir William Rothenstein, the principal. She had passed the test, a life drawing, and was assigned to the Design School under Professor E.W. Tristram. Eleanor Haley, who became her closest friend at college, had done a better drawing and entered the painting school, which was accorded higher status despite the presence on the design staff of such eminent tutors as Barnett Freedman, Eric Ravilious, and, from 1934, John Nash. Her diary for the next day reads, 'RCA design all a.m. then life and book illustration' and, on Friday, 'RCA architecture, V&A lecture and life'. Over the weekend she moved from a hotel to digs in South Kensington and on Monday writes, 'RCA all day, tea in CR', the Common Room in Queen's Gate: centre of student life for lunch, tea, long talks as well as the informal dances on Friday evenings. This was her pattern of life until her ARCA Diploma in 1936.

Skem continues, '"Life"', which meant drawing, drawing and yet more drawing of flowers, still life and the nude, lasted throughout the entire period of study. Of all the tutors at the Royal College, she spoke most often of John Nash and greatly admired his work. As for "design", this ranged widely to include lettering, lithography, hand-block textile and silk-screen printing and, in the third year, typography, with a session in the V&A museum or library several times a week'.

As if all this were not enough, in October 1934 she wrote in her diary, 'Went to Central School, took 1/4 hr. to find engraving room. Home 9.45'. This was the start of her wood-engraving course, which she pursued for the next three years on Monday evenings at the Central School of Arts & Crafts in Holborn, under Noel Rooke. A star exhibit of her diploma show were four full-page wood-engraved illustrations to *Gulliver's Travels*. Even as an undergraduate student, she began to build up a good collection of wood-engraving books. She enrolled for yet another evening course at the Central School on bookbinding, taught by William Matthews, which she did until 1937, and these two courses were to stand her in good stead much later.

Another of Nancy's interests was amateur dramatics. Many of her Scarborough friends were 'am-drams'. She cut a very striking figure playing the lead part in Bernard Shaw's *St. Joan*. An annual event on the social calendar

Nancy (right) at the Chelsea Arts Ball.

of Nancy and her circle was the Chelsea Arts Ball – a glamorous occasion.

Another favourite haunt of Skem and Nancy and their group of engineer and artist friends was the Blue Cockatoo restaurant, on the corner of Oakley Street and the Chelsea Embankment, where a bronze boy now swims with a dolphin that arches over a stone plinth. (The restaurant was bombed in the Blitz.) Skem tells me they could get a good dinner there in congenial surroundings for 2/6d. Jas Pirie remembers a very tall lanky waiter by the name of Oscar. One fateful day pigeon pie was served and enquiries were made as to the origin of the birds, whereupon Oscar lugubriously pointed a bony finger to flocks perched on the warehouses on the opposite side of the river, which perhaps accounts for the very reasonable prices.

It was during his student days that Skem developed an interest in classical music. He replaced his jazz drums with the recorder and his Duke Ellington records with Handel. Music had played no part in his upbringing. Beatrice was totally unmusical (though I do remember her liking the sound of a brass band). Skem educated himself in music entirely through the wind-up gramophone and he built up a collection of black shellac EMI records. (He never, at this stage in his life, went to concerts.) The discovery of Bach's music was an absolute revelation for him.

At the end of his three-year course, in 1935, thanks to his brilliance, eventual hard work, and fascination with the subjects he was studying, Skem graduated with a first, and was invited to stay on at Imperial College to do a PhD. Beatrice would have found this impossible to support financially. As the college had evolved from the City and Guilds of London Institute, it had close links with the City livery companies, who, as part of their patronage, endowed bursaries. Pippard recommended Skem for a Goldsmith's bursary. Perhaps remembering Mr Butler of the works and his job offer, his mother Beatrice apparently needed to be sure this was the right course of action and she made an appointment for a meeting with the Dean of Faculty. As Skem describes it, it was Beatrice who interviewed the Dean, rather than vice versa. In any case, the bursary was awarded, which covered his fees of £150 per annum and his rent of £2 per week and this enabled Skem to launch his research career. Perhaps unconsciously following in his grandfather Hector's footsteps, but now with his own fascination with the material, he started work on the study of reinforced concrete. However, after only a year, long before he got as far as submitting his thesis, a post arose at the Building Research Station (BRS), a government establishment at Garston, near Watford, in Hertfordshire, and Pippard, ever the mentor and encourager, suggested he take it immediately. Skem's salary, when he was appointed, was to be the princely sum of £240 per annum. In the years 1935–36, there were two million unemployed, and workers marched all the way from Jarrow to London to protest at their low wages. The economy was such that even the brightest graduates would *pay* an employer that much to take them on, so he regards himself as fortunate indeed to have been made this offer. With no hesitation, he wrote up the work he had done so far on concrete for an MSc and made the move to Watford.

During that summer of 1936, in the middle of the move, Beatrice and Skem had a summer holiday in Cadgwith, Cornwall, a tiny harbour village in a remote cove on the Lizard peninsula, with four or five cottages and a pub. There they met the artist, John Tunnard, who, as well as painting, was studying the local entomology and playing the drums in a band made up of fishermen. Skem and Tunnard immediately became friends. 'Anyone who met him was his friend in five minutes,' says Skem. Both he and his wife Mary (Bob), who was making hand-blocked silk at their cottage, had been at the Royal College of Art, but before Nancy's day. But more of Tunnard later.

Wood engraving of Cadgwith from The Wood Engravings of Mary Skempton.

Chapter 3

Building Research Station
1936–45

'L' eau dont les effets sont si subtils qu'il n'y a qu'une extrême droiture d'esprit qui y puisse y aller.' (Water has effects so subtle that only an extreme uprightness of mind can approach it.) *Pascal*

Orchard Drive, Watford

So, in that summer of 1936, Beatrice somewhat reluctantly uprooted herself from her many friends in Northampton and rented a house in Orchard Drive, Watford, Hertfordshire, on the new Cassiobury estate, in order to keep house while Skem started his new job at the Building Research Station at Garston. She told the story of house-hunting with compass in hand to ensure against a gloomy aspect. The house faces southwest and is full of sunshine. It is a typical 1930s bow-fronted semi-detached pebbledash house with a brick arch over the porch, French windows from the dining room into the long back garden, a kitchen with a white-tiled walk-in larder and a scullery, four bedrooms, inside and outside WC, and garage. Alec Clifton-Taylor in his television series on English towns, remarked of suburban pebbledash, 'Does one eat it, or has one eaten it already?' I only remember that, when I bounced my tennis ball against it in games of catch, pebbles cascaded down onto the driveway, leaving little bald patches below the bathroom window. Beatrice hung the family embroidered samplers and map in the sitting room, and the daily routine of the next three years began.

BRS

While many of his contemporaries travelled to Spain to join the International Brigade fighting General Franco, every day Skem cycled the few miles from the new house in Watford to the Building Research Station at Bucknalls House, Garston. BRS had been set up by the Government's Department of Scientific and Industrial Research. The DSIR was publicly funded and was not allowed to carry out paid consultancy. Skem was appointed on the Civil Service grade of Scientific Officer.

When he arrived, the director of the station was Sir Reginald Stradling, who incidentally was married to Pippard's sister. There were separate divisions: Physics; Chemistry; Architecture and Acoustics, and Engineering, divided into Structures and Soil Mechanics. Fred (later Sir Frederick) Lea was at that time in Chemistry. He was a great authority on the chemistry of cement. The head of the Engineering Division was Dr. (later Sir) William Glanville, but he soon left to become director of the Road Research Laboratory. He was succeeded by Norman Davey, a quiet scholarly man, interested in the properties of building materials. (He wrote *A History of Building Materials*, 1961.) The historical aspect of his work captured Skem's imagination and, in fact, the contact Skem made with people at BRS was the start of his lifelong fascination with the history of civil engineering, and the work on iron frames, canals, and the biographical work that so dominated the later stages of his career. Another influential person in this respect was Dr. S.B. Hamilton, a structural engineer, who was drafted into BRS during the war. His historical interests covered the whole field of the history of engineering, he wrote a paper on the 18th-century engineer, John Perry, and he was the first person to pay attention to the structures of mills. He later became President of the Newcomen Society for the Study of the History of Engineering and Technology, which was to be an important part of Skem's life. Skem's friendship with Hope Bagenal of the Architecture section was another factor that influenced his budding historical interest. Bagenal had been librarian of the Architectural Association in Bedford Square and author of a book on the history of construction.

When Skem first arrived at BRS, he spent three or four months in the Concrete section. He was put onto testing concrete beams and slabs and was given the 'ghastly task' of translating a German textbook on the theory of slabs. In order to undertake this, he went to German evening classes in Watford. However, Skem soon spotted the Soil Mechanics lab, and became fascinated by the 'curious bits of apparatus', the soil-testing equipment at the station, which had been built by C.F. Jenkin in the early 30s. Geology had been Skem's favourite subject as an undergraduate and 'my liking for geology was tied up with a liking for natural science. One was dealing with natural materials.' So he had a word with Dr. Stradling and asked if he could transfer from Concrete to the Soil Mechanics section. 'He said go there for a few months and see how you get on, and, of course, I never left.'

To come to the Building Research Station in 1936 was particularly exciting, as it was, at that time, the only place in England where the new science of soil mechanics was being developed. Other centres where soil mechanics was beginning to be studied were Harvard, where the Austrian Karl Terzaghi worked, and the Massachusetts Institute of Technology (MIT) with Donald Taylor. There was also a centre in Paris, where Armand Mayer ran a small Laboratoire des Batiments Publique, with perhaps one assistant, and in Delft,

Outside the BRS lab. Back row: Chaplin, Bond, Brown, Samuels and Smith. Front row: Meyerhof, Ward, Cooling and Skempton.

where T.K. Huizinga was doing government work in a lab linked to the university.

Skem describes the operation of the Soil Mechanics section at BRS in 1936–7:

> The railway companies would come to BRS and say they had a slip in the London clay north of the Kensal Green tunnel, for example. L.F. Cooling, head of the Soil Mechanics group, would say 'Oh, we would be very happy to investigate and advise, provided that we charge no fees.' This was absolutely typical of the way Cooling operated. In order to be completely independent, reports were submitted but no fees would be charged. I doubt whether BRS charged a penny for our work on Chingford. It was all done as research, but on practical problems.

(This was in any case required under Civil Service regulations.)

One of Skem's first jobs was on the foundations of Waterloo Bridge but, almost immediately, there was the event in July 1937 that launched Skem into his future career and fame, the failure of the earth dam at Chingford, Essex, which fortunately occurred *before* it was filled with water. But first, some soil mechanics background.

Karl Terzaghi

Terzaghi is indisputably the father of soil mechanics. He had spent the First

World War years lecturing on roads and foundations in Constantinople and had developed a passionate interest in geology. Richard Goodman in his biography of Terzaghi (Goodman, 1999), describes his realization that pure geological descriptions were inadequate for making engineering decisions, and his search for an understanding of how adequately to describe the behaviour of soils under engineering loads. Terzaghi had found that, previously, 'Theories were postulated without adequate presentation of the evidence on which their validity could be asserted. Failures were attributed not to errors in the postulated basis for design as much as to the perverseness of nature.' He had realized that, 'To treat soils as engineering materials would require knowing how they deform under different kinds of loads. Expressing this quantitatively would require developing apparatus for soil tests and advancing theories to convert the results into values of fundamental soil properties. Terzaghi decided to begin a methodical investigation in this vein, starting with sands, one of the most important components among the variety of soils in nature.'

Devising primitive equipment (forerunner to the oedometer, an instrument for measuring the compressibility of soil) in his lab at Robert College, Constantinople, and later at MIT and Harvard, Terzaghi studied the forces acting between the grains of uncemented sediments. 'These forces stem from pressure of self weight, and chemical action of water in contact with the grain surfaces' (Goodman, 1999). 'He was well aware that the properties of the soil mass would depend heavily on the proportion of void space in the specimen and the degree to which the pores were filled with water. The void proportion, in turn, was obviously increased or reduced with change in the magnitude of the inter-granular forces... He sketched a fundamental apparatus in which he could both measure permeability and monitor the pore volume change on changing the external pressure on a specimen.' This work led him to understand that 'the addition of an increment of external pressure to a clay resulted in a temporary increment of water pressure of equal magnitude. The soil's inter-grain contacts would feel no increment of force until the pore pressure began to dissipate. When the water pressure in the soil's pores had a value u, and the external pressure had a magnitude p, only the value p-u was effective in causing force between the grains. He called this quantity "the pressure acting in the solid phase of the clay". Today it is known as the effective stress'.

He had published his seminal book explaining this and other principles, *Erdbaumechanik*, translated in the English-speaking world as *Soil Mechanics*, in 1925.

At Harvard Terzaghi worked with Arthur Casagrande, whose brother Leo had been brought from Germany to BRS immediately after the war to describe and continue his work with electro-osmosis. He stayed there for about two years, before moving on to join Arthur's engineering consultancy in the USA.

Skem heard about *Erdbaumechanik* during his early days at BRS (and reread it in the 1980s when he was working on his paper, 'The History of Soil Properties') but was also very much influenced by the numerous articles on effective stress that appeared in American journals, and therefore in English. (He had access to American publications by the good offices of a Boston friend of his mother's, Arthur Shaw, partner in the engineering firm of Metcalf & Eddy. He was a member of the American Society of Civil Engineers (ASCE) and sent 'Junior' papers of interest emanating from Harvard and MIT.)

The influence of this work led to the setting-up of the First International Conference on Soil Mechanics at Harvard in 1936, at which Terzaghi was a leading figure. Only a few hundred of the proceedings of this important conference were produced. Skem saw Cooling's copy and managed to get hold of one of the last copies for himself. He explained the basic idea of effective stress to me in this way: 'The vertical load is trying to push soil particles together, while water between the particles is trying to hold them apart.' He also explained that, before Terzaghi enunciated the principle, people couldn't imagine that water was working in this way on what are minute particles that you can only see under a microscope. For example, with clay, which was seen as a solid object, 'You can make a bore-hole in London clay and not find any water, but come back four days later and there it is.'

Terzaghi was a handsome and dynamic figure with a Viennese accent, who dominated any gathering. He held strong opinions on very many things, which, particularly later in life, he did not shrink from expressing forcibly and colourfully. The extent of Skem's admiration for Terzaghi is expressed in his 1950 BBC broadcast.

> Great mental courage was required to see that… the apparently intractable problems of soil mechanics could be placed on a sound scientific basis and, at the same time, in a form useful to civil engineers. In 1920 such courage was applied to these problems by Dr Terzaghi, for long Professor at Vienna and since 1939 at Harvard. Terzaghi has been largely responsible for the immense developments in the subject which have taken place in the past 25 years. He has a directness of mind which is met with in only a few individuals of any generation, and by his own research, and practice as a consultant, and by the work of men inspired by him, the 'intractable problems' have been successfully tackled. The subject is now of a real value to the civil engineer and is also contributing to our knowledge of the physics of materials.

The Chingford dam

If it had happened today, the Chingford dam failure would have been an item on the *News at Ten* and made the front page of *New Civil Engineer*. Dr Arthur Penman, who worked at BRS from 1944 and is its unofficial historian, wrote the following account of the failure and its technicalities: 'The Metropolitan Water Board (MWB) had their own design team and had built several reservoirs on the relatively flat land adjoining the Thames and Lea rivers, by the construction of encircling dams. A contract had been let to Mowlems for a dam near Chingford in the Lea valley, adjacent to the existing twenty-five-year-old (and eight metres high) King George V reservoir. During construction in 1937, a failure occurred and Jonathan Davidson (later Sir Jonathan), Chief Engineer to MWB, who served on the BRS Board, sought advice, resulting in an investigation by the Soil Mechanics section. As with the older, adjacent dam, construction was over a layer of soft yellow clay that covered the flood plain of the Lea. Excavation revealed a slip surface passing through the soft clay and up through the puddled clay core, as indicated by the drawing. At the time of the slip, bank fill had been in place for only eleven weeks. The shear strengths of the clays were measured by unconfined compression tests on undisturbed samples. A new two-circle phi=0 analysis was developed to calculate a factor of safety of unity. The question was then raised as to why the King George Dam had been successful. Skem explained this by the different speeds of construction. The older dam had been built twenty-five years earlier, in a slow manner, but Mowlems had brought in Caterpillar D8 tractors pulling tracked Athey tipping wagons for the new construction, which had consequently been much more rapid. The fill level had been raised four metres during the month prior to the failure.' There had not been time, Skem argued, for the layer of soft clay to be consolidated by the weight of the new fill and gain sufficient strength. Using consolidation theory given by Terzaghi and Frohlich (1936), he calculated the

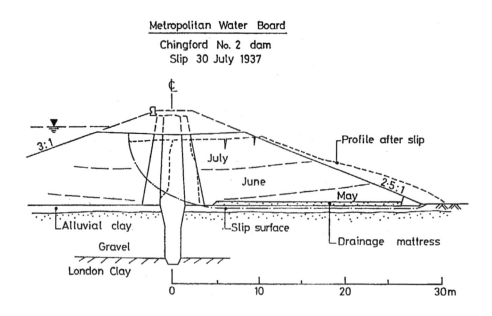

Section drawing of Chingford dam, from the contemporary account.

pore pressure in middle layer of soft yellow clay under the weight of the overlying bank fill immediately after it was placed and also eight months later, when significant consolidation had taken place. By means of consolidated, or 'equilibrium' shear-box tests, he predicted the gain in strength during this consolidation and the predicted value compared favourably with values that were measured in the field eight months after cessation of fill placement.

Skem points out that the BRS group who were called in by MWB to investigate the failure was an untried group. He himself was a mere 23 years old, Cooling was basically a physicist, and this was their first big job. They were at the slip within two days of the failure and, on this first visit, it was still moving slowly. Skem remembers they had a little trench dug down at the toe of the slip and he 'put two nails in the clay, one above and one below the slip surface, and a few hours later they were perceptibly further apart'.

In an interview in 1998, Dick Chandler of Imperial College asked Skem about the tests they had carried out:

Chandler: Were you doing oedometer tests in those days?

Skem: Yes, we would have done, to find the rate of consolidation of the clay. And shear-box tests. What we called immediate and equilibrium tests in those days. Equilibrium was nearly drained but not quite, because we did them too quickly. We consolidated them fully, but then sheared them rather too quickly to be properly drained.

Chandler: The immediate ones were not consolidated at all?

That's right. Indeed, chiefly these were unconfined compression tests, although I did do one or two undrained box tests with solid ends, out of interest. Using non-permeable end-plates.

Chandler: But you would normally have used porous stones?

Skem: Yes.

Chandler: And at least attempted to consolidate, that was your intention?

Skem: The so-called equilibrium tests were consolidated for a day or two days, and then we would shear them in about three or four hours. They were very slow, but not quite slow enough! And, in fact, I think the results were probably within a degree or two of the correct answer.

Chandler: And the intention was to have some numbers to feed into a back analysis of the failure?

Skem: Yes, that's right. And it was Golder who devised the non-circular analysis. By using circles of two different radii. Two arcs, one of which could have an infinite radius if you put a straight line through it... He worked that system out, and tried different positions of the interacting force between the two. My contribution was the consolidation aspects of it, proving how the strengths had increased during construction.

Chandler: As the result of some consolidation.

Skem: Yes, quite a bit actually, but not sufficient to maintain stability.

Chandler: Which would have been the case twenty years earlier with...

Skem: Absolutely, because Chingford was the first big job in England with modern American earth-moving machinery, so the section went up quite quickly and, needless to say, it was the highest part that failed.

Chandler: It was the same design as the earlier dams?

Skem: Yes, the King George V reservoir was *immediately* next door. I mean there was literally only a small road between them, and so we had the slip on one side of the road, and the old bank, four feet higher than Chingford was at the time of the slip, just across the road, which had been perfectly stable.

Chandler: All very puzzling for everybody!

Skem: It was very puzzling to the engineers, yes.

Chandler: And how long was it puzzling for you Skem?

Skem: No time at all. We spotted almost immediately that it was a consolidation problem. The King George had been built in, I'm going to say, 1910, with horses and carts, and it had taken them nine months to build the bank instead of three months at our site!

Chandler: So you were obviously more than familiar with what one might call the modern ideas of consolidation, and also the fact that that resulted in a gain of strength?

Skem: Oh, yes. There was nothing novel in that. The novelty was our application to this particular engineering problem. Consolidation was usually thought of in terms of building settlement... I am sure we made it clear that we didn't think that the Metropolitan Water Board was to blame, because the cause of the failure had been this consolidation phenomenon. Although I have just said that the phenomenon was well understood, that was only within the Soil Mechanics section at BRS.

Chandler: Not in the sense of being incorporated in a British Standard.

Skem: Oh, no! Not remotely!

Chandler: Just the small group of you, and perhaps one or two other people in the country, if that?

Skem: I'm trying to think who else, but I think there weren't any in England. Possibly someone like Fisher Cassie, but I doubt it. I think there were just the three of us who knew what was going on. Anyway, the Water Board then asked us to redesign, which we did, but the Board was not very happy about it. The costs were huge because of the extra volume involved. The dam was three to four miles in length. So an impasse was reached, which lasted some months. Robert Wynne-Edwards was Mowlem's agent on the job (by this time we had got extremely friendly with him, he was a very good chap) and had taught himself what we were doing, he understood what we were up to.

Their scheme was carrying no weight with MWB because of the lack of status of this young group. Skem continues, 'I remember Wynne-Edwards coming to BRS and saying, "Look, we have to get the absolutely top man in on this job, no matter at what cost. Who do we go to?" So we said Terzaghi, who was at that time in Paris. Wynne-Edwards took one of the early Imperial

Working on the Chingford Dam failure, 1937. From the left: Hugh Golder, Skem, unknown, Robert Wynne-Edwards and Leonard Cooling.

Airways flights from Croydon Airport to Le Bourget to persuade Terzaghi to come to give independent advice.'

Goodman takes up the Paris story. 'The affable blond blue-eyed Wynne-Edwards solemnly spread out plans, profiles and borehole data. After inspecting them, Terzaghi asked where the dam was located and was told it lay fourteen miles north of London Bridge. He then said: "That dam must have been designed by an enemy of the British nation, because it will fail, whereupon your Parliament and Westminster Abbey may be washed into the Thames." Now [Wynne-Edwards] smiled as he reported that it had already failed. Terzaghi asked what instructions he had received from his boss. The reply: "To watch your face while you look at the profiles. If you don't show any signs of disapproval, I should take my hat and go. If you are shocked, I should put you into an airplane and bring you over."' (This account is taken from Terzaghi's diary for April 22nd 1938.) Skem continues, 'To our enormous relief Terzaghi approved of our report. He worked very speedily and produced his own design, which, needless to say, was vastly more costly than anything we had suggested... He did do some calculations, but I think they were only a gesture, he just put in a huge gravel key. The work was carried out exactly in accordance with his report, and at vast cost!'

To Skem, the young engineer on his first big job, Terzaghi appeared as a heroic figure. He seemed charismatic, and Skem ever after regarded him as somewhat of a mentor in his professional development.

Terzaghi very much enjoyed his trip to London. Wynne-Edwards entertained him in the best gentlemen's clubs in London, and in his diary Terzaghi describes with relish a lavish dinner of caviar and asparagus. (He also conveys the excitement and exhilaration of those very early days of commercial flying – describing sunset over the clouds on his way back to Le Bourget.)

In his biography of Terzaghi, Goodman describes a somewhat different version of events. He writes, 'The English engineers proposed stabilising the dam by flattening the slopes, but Terzaghi showed that the failure would occur regardless of the embankment's sideslopes because the fill was simply being piled too high for the strength of the underlying yellow clay. It hardly mattered what inclination its sideslopes were given; the foundation would still fail.' As Penman comments, 'This flies in the face of Skempton's detailed calculation of the expected strength increase which was proved by actual site measurements of the strength of the clays 37 days after failure, and again after 277 days. The suggestion that flattening the slopes would not improve stability is contrary to general geotechnical experience, and must be regarded as a mistake.' (personal communication)

Terzaghi's report to Mowlem's of May 21st 1938 even implies that the embankments of the older King George reservoir are also unsafe, but in this also he is mistaken. Indeed, Cooling later had to put him right on a number of points relating to Chingford (letter 6.7.1943). Goodman also quotes a letter from Terzaghi to Casagrande speaking of the 'appalling blunders' made at Chingford (letter 8.6.1938). In fact, perusal of the letter itself makes it clear that this typical Terzaghi overstatement refers to the MWB's original design, *not* the BRS group's suggestions. (He states: 'I found the following situation. The project of the Water Board was as silly as it could be. Never, outside of Soviet Russia, have I encountered such appalling blunders. Cooling of the Research Station made a very reasonable report on the situation but he lacks both the experience and the authority to impress the Water Board ... Hence, it was a splendid opportunity to show what we of the Soil Mechanics camp can do and I took full advantage of the situation. The resistance against my revised project was practically nil and the Research Station was delighted to be backed up in its bashful efforts.')

Terzaghi showed great affection and respect for Wynne-Edwards personally and for the group as a whole and Skem in particular, as demonstrated in so much subsequent correspondence. Skem's clear memory of the events at Chingford is that Terzaghi backed up his own conclusions, giving him credibility with the weight of the Water Board. Terzaghi later somewhat distorted the events of Chingford to his own advantage, for example, in a letter to Ackerman of February 4th 1942, he took complete credit for saving the dam. This is understandable in the context that he, an Austrian, was seeking to be given engineering work in a country where, at a time of war, he was presumably viewed as an enemy alien.

Penman again: 'This case aroused a considerable interest in soil mechanics and the laboratory at BRS within the civil engineering profession. Universities became interested in teaching this new subject and heads of the civil engineering departments began going to BRS to find out more about the subject.'

Silas Glossop

Wynne-Edwards' sub-agent from 1938 was Rudolf Glossop, known to all as Silas. He was 12 years older than Skem. He found it expedient to set up a small soil-mechanics laboratory on the Chingford site to take and test samples during continued construction to the modified design. This was the very early beginnings of a commercial laboratory, which later became Soil Mechanics Ltd. During the rebuilding of the dam in 1938–39, and for the rest of their lives, Glossop was one of Skem's greatest friends. They were both giants, physically as well as intellectually. Glossop's background was very different from Skem's. He was from a well-to-do Derbyshire family whose money came from involvement in lead mining, and his father was a bank manager at Bakewell. He had worked as a mining engineer in Canada and then in the Gold Coast in the 1930s, before joining Mowlem's, initially under Sir Harold Harding. As well as his considerable physical presence, Silas had a sense of his own grandeur and there was an aspect of him that was 'a showman', in his wife Sheila's phrase. Bill Allen (a colleague in the acoustics section at BRS) tells the story that, when Glossop was in charge of Mowlem's Thames Division, he hired an admiral's barge to take him up and down the river. This was typical of his style. Skem says appreciatively of him, 'He was a starter, a doer, an entrepreneur.'

The Guggenheim Jeune

Just before they met over the Chingford failure, Skem and Silas had coincidentally met in quite another context. Skem went to a private view in March 1939 at Peggy Guggenheim's gallery, Guggenheim Jeune, at 30 Cork Street. She was holding a one-man exhibition for John Tunnard, whom Skem had met while on holiday in Cornwall with Beatrice

Silas Glossop as a young man.
Photograph by Frank Freeman.

the previous summer. Skem bought one of his watercolours for £5. A girlfriend of Silas' told him about this 'remarkable young man who should interest you. He is doing research on the properties of clay, and he bought a picture from John Tunnard's exhibition at Peggy Guggenheim's gallery'. Silas thought, This is someone I must meet. The girlfriend had forgotten the name, so Silas sped to Cork St to find out who this was. In the absence of Mrs Guggenheim, an assistant was in charge. 'Can you tell me the name of a young man who bought a Tunnard and is researching on clay?' enquired Silas. After some confusion, it eventually dawned on him that the assistant thought, not unnaturally, that he was talking about the Bauhaus artist, Paul Klee (pronounced 'klay')! This is the kind of joke that would appeal enormously to Skem and Silas.

Shortly after the private view, Silas visited Skem at BRS and, as a result of their contact, he became passionately devoted to soil mechanics, teaching himself about it at high speed. Wynne-Edwards, who was by this time on very friendly terms with Terzaghi, his 'Dear Baer', wrote him an entertaining letter in July 1939, telling of his difficulties in persuading George Burt (Director of Mowlem's) of the need for an on-site testing lab at Chingford, and extolling Glossop's potential. Glossop's lab at Chingford was, in fact, very successful and was soon being commissioned for other consulting jobs, for Mowlem's and others. During the 1939–45 war Glossop worked on airfield construction, but took his lab and its technician, Terry Clark, with him. By 1943 the temporary laboratory had become Soil Mechanics Ltd. The firm was located at 123, Victoria Street, SW1 (in Sir George Burt's flat). Silas Glossop managed it together with Harold Harding and Hugh Golder, who had the previous year himself recruited Bill Ward to replace him at BRS so that he could leave government service and join in forming this first British commercial soil-mechanics laboratory.

One reason for Skem's admiration for Silas was that he always read enormously and widely, especially philosophy, (he knew Bertrand Russell). He also wrote poetry had been part of the sophisticated world of artists in Paris in the 1920s, and had had an affair with the beautiful American Djuna Barnes (who had also been the lover of T.S. Eliot). He had had a relationship with the novelist Antonia White, mother of his daughter, Susan Chitty, who later wrote a biography of Antonia in which Silas comes over as a reliable and concerned father. Silas and his girlfriend Betty Butterworth were together on and off for about five years and she inspired his poetry but it was never on the cards that this would be a permanent relationship. She had her own flat in Chelsea, near the Albert Bridge. During the war, while he was building airfields, Silas based himself in a cottage in the New Forest, where Skem and Nancy visited him and Betty. Susan was also there as a child and Nancy gave her riding lessons in the forest.

Although Silas' behaviour with women seems to have been very different from Skem's, their humour was similar and they had many interests in common. Skem valued Silas' involvement in the artistic and literary scene and his contacts with industry but, more than that, he valued Silas' generous larger-than-life personality and intelligence, and devotion to the cause of geotechnical engineering. In return, Silas valued Skem's capacity for application and perseverance and his academic ability. (Skem later gave Silas considerable help with his 1968 Rankine lecture.) When they were together, they talked and talked until late into the night, in clouds of pipe smoke and an atmosphere of mutual stimulation of ideas and laughter. While they were working on the Chingford job, Skem and Silas decided to share a top-floor rented flat in Oakley Street, Chelsea, no more than 100 yards from the Blue Cockatoo restaurant (and convenient for Betty Butterworth's flat). It was a *pied-à-terre* in London for both of them. The rent was 27/6d per week, of which they paid half each. Silas and he were sometimes at the flat together, but more often came at different times. The flat must have been a wonderful escape from the loving but powerful maternal presence of Beatrice in Watford and it must have been during this period that Skem took Nona Eggington to the Cab Calloway concert. The house was bombed during the war and there is now a block of flats on the site.

Bucknalls

The engineering labs and offices at BRS were in the stable block and

The early BRS buildings. Photograph by Arthur Penman.

outbuildings of a large Victorian Tudor-style house called Bucknalls. Penman described to me the layout as it was in the early 40s. 'There was a south-facing garden wall, on to which, years ago, a lean-to had been built that became the soils lab. It formed a corridor of offices. Skem had the office at the end, which occupied the full width of the lean-to. He had a desk, a big bench, a sink with a tap, and in the floor was a piece of concrete, replacing the suspended floor, to support the weight of the four oedometers which were also in the room. Next to him was the open lab, and in the corner just outside his door there was a fume cupboard, against the whitewashed garden wall, and a solid slate bench, where Brown and Bond used to do liquid limits tests. Skem had written on the whitewashed wall outside his office, 'Patience is necessary with clay, as with women. Soil mechanics proverb, early XXth cent.' Later, visiting students from across the world translated this tag into Chinese and Arabic! Bill Allen remembered that Skem's room looked out onto an old mulberry tree in a courtyard between buildings and that, in his last year at BRS, Skem had hung an interesting painting by Tunnard in his room. (This is the oil that now hangs in Skem's room at IC.)

Hugh Golder tended to the poetic in his spare time, and wrote the following amusing doggerel about these BRS days:

> Once Garston raised a brood of bastard chicks
> Who had no thought of steel or stone or bricks
> But, fathered by some unknown strange gargoyle,
> Began to play about with pats of soil,
> And working in a humble glass-lined annexe
> Were metamorphosed into Soil Mechanics.

In 1938, Pippard invited Terzaghi to come to Imperial College, but 'Terz' decided instead to return to Harvard where, in the meantime, his former assistant, Arthur Casagrande, had established himself. Before leaving, however, he delivered, at Pippard's invitation, a seminal lecture, the 1939 James Forrest lecture at the Institution of Civil Engineers in Great George Street. In this lecture, entitled 'Soil Mechanics – A New Chapter in Engineering Science', he discussed the contributions of soil mechanics to the calculations of the stability of slopes, and the influence of pore pressures on the shearing resistance of clay. Skem says, 'The lecture was a milestone. It was fully attended by lots of senior people. Terzaghi was immensely impressive. He carried a colossal presence. He dealt with a number of problems very simply, no-one had ever heard of these things before. It was all set out very simply and very clearly by someone with obvious authority. He was a magnetic personality, an absolutely extraordinary chap. The James Forrest lecture put soil mechanics firmly on the map for most British civil engineers' (Interview by Dick Chandler, June 1998).

As the eminent American civil engineer, Ralph Peck, pointed out (speech at the 11th Int. Conf. Soil Mech. in San Francisco 1985) at the time of the

Tunnard's painting 'Circle phi' combining elements of the flood relief channel in the Fens and the Fellenius analysis. From the colour photograph by Colin Crisford.

James Forrest lecture, 'the determination of the shearing resistance (of typical clays) required a knowledge of the pore pressure in order to obtain the effective pressures, and there were no means of forecasting the pore pressures under field conditions.' This was the problem to which Skem was to address himself after the war.

It must have been on this visit to England by Terzaghi that a story told by Bill Allen unfolded. The famous American architect, Frank Lloyd Wright was in England at the same time, giving a series of lectures at the Architectural Association, on the Tuesday and Thursday of two consecutive weeks. On the Wednesday of the second week, Bill Allen invited Lloyd Wright down to see BRS. That lunchtime, Terzaghi was looking out of Skem's window into the courtyard and saw a figure emerging from a car. He said, 'Is that really Frank Lloyd Wright I can see getting out of that car? I come halfway around the world only to meet FLW, who lives a few blocks away from me in Chicago, and I have never met him there. What is this BRS, the centre of the civilized world?' A memorable lunch was eaten, hosted by I.G. Evans (who had by then succeeded Stradling as Director of the BRS), with Frank Lloyd Wright, Terzaghi, Cooling, Bill Allen, and Skem. (At the time Terzaghi was working on the Chicago subway, with Ralph Peck as his young assistant.) Skem remembers that, at this lunch, Lloyd Wright was boasting about a hotel in Tokyo that he claimed to have designed to withstand earthquakes. It had indeed survived the great earthquake of 1926 and he claimed that he had calculated the settlement of the building in advance of its construction. At this, Terzaghi gave an ironic smile because at that time he himself was the only person who had the expertise to do such a thing.

In the last hot summer before the war, Skem and Nancy had a holiday in St Tropez, on the then relatively undiscovered French Riviera, during which Nancy did the sketches for her sun-filled paintings of St Tropez harbour.

War work and marriage to Nancy

When war broke out in 1939, Skem, as a government research engineer, was in a reserved occupation. He had no guilt about not going into the forces. He felt his family had more than paid its dues to the country in the suffering undergone by his father in the First World War and the subsequent failure of the State to recompense or support Beatrice with a war pension. Skem now says that 'a large number of people were on reserved occupations, not because these might contribute to the war effort, although it was possible that they would, but because it was felt so necessary to have the intellectual infrastructure in place for when the war ended. It was a perfectly conscious policy and an example of remarkable foresight, and we thought it was remarkable at the time.'

The fen drainage scheme was an example. 'It was urgently needed, and so the site investigations were put in place, which had absolutely nothing to do with the war. The idea was, as far as possible, let's look forward and plan for the peace, and have the people to do it. Dunkirk was a low point, but after Hitler did that lunatic thing of attacking Russia (everyone remembered Napoleon), that he would be defeated was a foregone conclusion. So, although a lot of what we did at BRS was to do with military things, by no means all of it.' (Interview with Dick Chandler, June 1998).

For one of his war-related tasks Skem was seconded to the Ministry of Works, and sent to southwest England, where, from October to December 1939, he was to oversee the manufacture of reinforced concrete air-raid shelters at Barnstaple, Plymouth, and St Austell. Initially, many of the shelters did not come up to standard, especially the Plymouth ones. Nancy went with him to Cornwall, and they took digs in the village of Mevagissey, overlooking the harbour. Nancy did two beautiful paintings and several wood engravings of the view over the harbour at Mevagissey. Jas Pirie came to spend a weekend with them and they stayed together at the Western Hotel in St Ives and had a sunlit picnic on the cliffs. One great advantage of being in Cornwall is that they could keep up Skem's local connections. They visited John Tunnard and his wife, Bob, who were by this time running a hand-blocked silk-printing business in Cadgwith.

In January 1940, Skem came back to BRS and was put to work on naval dockyards. He was promoted to Senior Scientific Officer, on £400 a year. This meant that he could, at last, afford to get married so, in June, he and Nancy, with only Frank and Pamela Freeman as witnesses, went at short notice to Watford Register Office and on to the pub for some celebratory beers. Wanting the minimum of fuss, they did not even invite their parents to the wedding. Skem says he probably went back to work the same day! This rather low-key event probably reflected the grim times and the lack of any particular religious inclination on either Nancy's or his part, but it may also have indicated some of Skem's ambivalence. My mother told me years later that, in the true spirit of research, Skem couldn't easily make up his mind to marry one woman when he had not been able to explore fully all the available options! I have no doubt, however, that she had enough determination for both of them.

Skem and Nancy returned to Cornwall for a honeymoon, during which they joined John Tunnard in his fascinated study of the flora and entomology of Cornwall. Nancy studied and pressed wild grasses, and made sketches for her oils and engravings of Mevagissey.

Skem gained his AMICE in 1941. In those days (it is much more difficult now), this involved building up a body of experience and presentation of a research report. Skem submitted his investigation of the bearing capacity of soft clay soil. This arose out of his work on a foundation failure at Kippen, near Stirling. This was a very early full-scale check on bearing capacity of clay and became a classic for that reason.

Muirhead

Penman tells the story of Skem's next important BRS dam failure investigation, in 1941, at Muirhead on Rye Water, in Scotland. The engineer in charge there was Mr James Banks, senior partner in the firm of Babtie, Shaw & Morton. (He later became President of the Institution of Civil Engineers.) They had got the dam two-thirds built when great cracks appeared along the top and bulging at the bottom. Skem recognized that it was a rotational slip. Banks, loath to think this was happening, staunchly maintained, 'It's not a failure, only a wee movement!' Tactfully saying how much they would like to measure the 'movement', Skem and Golder put white marker pegs all the way down. The movement stopped, and Banks proposed that they let the contractor go ahead. Skem and Golder suggested waiting until they had the pegs surveyed, but allowed some fill to be put on, to demonstrate the theory that an amount of extra material would make a difference to the stability of the dam. Skem and Golder were spending most of their time at Muirhead, staying in digs at Largs, which Skem remembers was bleak and cold in winter, returning to BRS occasionally to do war work. Bill Allen tells of Skem's fury one morning when a phone call came through to BRS saying the contractor at Muirhead had put double the specified amount of soil on the dam. Bill, who had taken the call, rushed from his building to Skem's to tell him this news. Skem shot up to Scotland. The movement was very pronounced, all the pegs had shifted, and you could see exactly what was going on in the slip. Penman points out that, at the same time as Muirhead was failing, the construction of a sister embankment dam at Knockendon, also involving Babtie, Shaw & Morton, was a third of the way up. Skem advised measuring pore pressures and they put about six crude standpipes into the fill, two-inch galvanized water pipes, perforated at the lower ends. This was, he thinks, the first example of pore pressure measurement in a British dam. Among the remedial measures for both dams was the use of a berm to stabilise the upstream slope.

Cassio Road, Watford

Skem and Nancy set up home in a top-floor flat in Cassio Road, Watford, within cycling distance of Garston. The rent for this was 27/6d per week including rates. It overlooked the West Herts. cricket ground. This felt familiar to Skem from his memories of Abington Avenue, Northampton. An insight into Skem's new domestic contentment comes in a long letter to Silas Glossop, 'Nancy… is developing both her formal and colour sense…both of us are working very considerably better nowadays, and I am finding a moderately stable domestic life very suitable. I work about four evenings a week, and with a contented mind. If it weren't for the bombs and war background, I feel I should be getting into my really useful period. However, even as things are, I am enjoying life quite a lot…I long for the days when Chelsea is habitable again, and the Café Royal stays open later that 8 o'clock!' Guthlac Wilson (who had come over from working with Terzaghi at Harvard and was to become a good friend) went to stay at Cassio Road, and wrote to thank Nancy, 'It is nice to be amongst good books, good pictures, and of course a good bone to chew on (soil mechanics again!).'

The front door of the flat was never used. You came in through a gate in the garden fence and then had to climb up an iron fire escape to the kitchen

The women in Skem's life at Watford: Bea, Nancy, Martha, Olive with Judith at the front.

door. As a toddler, I found it terrifying looking through the metal patterns of the stair treads to the ground, way below. While Beatrice looked after me, their baby girl, Nancy gave private painting lessons to two local ladies and taught art at Henrietta Barnett, a Girls' Public Day School Trust school in North London, travelling there daily by bus.

At weekends, when Skem was not working, they cycled around the Hertfordshire countryside and surrounding Chiltern Hills and villages, with me in a baby seat on the back of Nancy's bike. She usually took her sketchpad, and turned her drawings into landscape oil paintings and beautiful wood engravings.

Nancy's art teaching did not last long, however, because all three of us went down with chickenpox, and Skem blamed the school as the origin of the infection. Nancy told me that he forbade her to carry on teaching in order to prevent a recurrence of childhood diseases in the family. The late 1940s was, in any case, a time when great numbers of women in England were encouraged to cease employment outside the home in order to care for their post-war 'bulge' children and make way for returning service men in the labour market.

Other embankment dams

Following the Chingford success, the Metropolitan Water Board had asked BRS to help with the design of their next big reservoir, the Walton (now Queen Elizabeth II). The Board was making a site investigation and a contractor was taking samples from the clay foundation at regular intervals along the very long centre line for the new embankment. These were sent to BRS for testing, to check that the foundation would be strong enough to support the proposed dam and that there would be no risk of a Chingford repeat. Arthur Penman was given the daunting task of testing all these samples. Alan Bishop, who had taken a job with the MWB after graduating from Cambridge, came to help the BRS group with this massive testing operation, bringing with him the autographic triaxial testing equipment he had designed and had built at the MWB workshops. This machine was set up at BRS and over 400 samples were tested from the Walton site.

Just at that time, the MWB had refilled the King George V reservoir after it had been drawn down during the war in case of bomb damage and it began to leak dangerously. Penman and Bishop investigated this condition, which they found was due to shrinkage and cracking of the upper part of the puddled clay core. Later, Skem pointed out that the clay for that core was an alluvial clay of very high plasticity. A puddle core of London clay with lower plasticity in yet another MWB dam, subject to the same water-level restrictions during the war years, had suffered no appreciable shrinkage.

Skem quickly built up a close relationship with Bishop, a slim, somewhat reserved Quaker. Skem admired his extremely quick intelligence and encouraged him to write up the Walton test investigations as part of his 'Part

C' for the Civils. He invited Bishop to stay at Cassio Road for a few weeks until he was able to find digs in Watford. Skem later wrote,

> He had quickly fitted in with the rest of us and, not long after his arrival, Cooling made an attempt to get him onto the BRS staff. This fell through, not because of any lack in Bishop's abilities, which were evidently of an exceptionally high order, but as a result of him being a Quaker. As a strict member of the Society of Friends he had been a conscientious objector during the war and this debarred him, in the eyes of the Civil Service Commissioners, from taking a post at BRS.

This is an indication of the suspicion with which pacifists were regarded, even after the war, by Government.

When he wasn't on his bike, Skem used to travel to and from BRS on the Number 321 bus, which ran close to the Watford flat. It stopped running at about 10 p.m. One night Bishop, Skem, and Penman were in the lab, working late on these triaxial tests and drawings, when suddenly Skem found it was something like half-past eleven and he had missed the last bus home. Penman, because he lodged with Sid Burns, the BRS caretaker, had access to the master key and said, 'Don't worry. I'll get one of the vans and drive you home.' He got out the little Morris ex-tea van and off they went. It was the blackout, and they had shades on the headlights. As they came to the Dome roundabout on the Watford Bypass, they were stopped by police doing some traffic check. Penman was perturbed and thought they might be in difficulties but Skem launched into a most eloquent tirade: 'Look here, my man, we are Government officers on important Government business. Would you mind standing to one side!' The policeman touched his cap, and they continued on their way.

Bill Allen remembered being on wartime fire-watching duty with Skem. They were responsible for a wooden warehouse storing food in North Watford. One Saturday, they had a night off. Next day, he and Skem were taking a walk in Cassiobury Park, and remarked to each other how nice it was to smell bonfire smoke in the spring. They soon realized, with horror, that it was the smell of the warehouse that had burned down the night before, not as a result of enemy action, but of an electrical fault. The mixture of baking flour and melting cheese made, Bill remarked, the largest cheese toasty he had known.

One day, Bill and Skem were sitting on a pile of gravel at BRS, eating their sandwiches, and Bill told him about how worried he was about becoming too narrowly specialized. He was surprised to hear Skem say he was worried about just the same issue. Skem was wondering if he shouldn't become a drummer or a professional flautist, or train as an architect. He was not sure if his work at BRS was valued. Bill was astonished that he should feel this and told him to go to Fred Lea, who was by then the Director of BRS, for reassurance as to his value.

The cultured engineer

Skem's discovery of the worlds of art and music was widening his social and cultural horizons. He began to move in more intellectual and aristocratic circles than his provincial Northamptonshire background had previously allowed. He and Nancy played recorders together in the early years of their marriage but he soon graduated to the flute and took lessons with John Francis (who was married to the harpsichordist, Millicent Silver, whom Nancy knew from Henrietta Barnett – they were teaching music and art, respectively).

Hope Bagenal became a close friend. He and his wife, Alison, were a generation older than Skem. Skem says of him that 'his was the most cultivated and scholarly mind I had ever met' and he 'never needed a peg to hang a story on'. The Bagenal family goes back to the time of Queen Elizabeth I. The Elizabethan ancestor was given hundreds of acres of ground some of which was still in the family. His sister, Faith, had been a fellow guest, with Lytton Strachey, at Lady Ottoline Morrell's Garsington. Hope's daughter, Kate, was a singer and Skem's age and it was she who later invited him to play the flute in Bach's *St Matthew Passion* at Bury St Edmunds. He also played seven times at the Leith Hill Music Festival in Dorking under Ralph Vaughan Williams. Skem and Nancy spent many weekends with Hope and Alison in their ancient and beautiful cottage, *Leaside*, in Hertingfordbury. Their romantic garden led down to a river. I remember Nancy paddling a canoe with me as a small child, and her great delight when she saw the azure blue flash of a kingfisher along the river bank. It was the trees there that in 1944, inspired one of her most accomplished and beautiful wood engravings, *Willows, Leaside*.

Nancy had done some wartime work as nursing auxiliary at Watford Hospital and there she had made friends with Joan Stokes, who worked in the pathology lab. She and her husband John, later a consultant at University College Hospital, were able amateur musicians. John was a good pianist. Another of this circle was Dermot McCarthy, son of the *New Statesman* and *Sunday Times* critic, Desmond McCarthy. He lived in a mill near Hemel Hempstead and had a very beautiful French wife. (She was awarded the Croix de Guerre for risking her life for accompanying wartime pilots who had survived crashes to English boats waiting at the French coast.) Dermot was a doctor at Great Ormond Street Hospital, which had moved out of London to an ex-workhouse in Hemel Hempstead for the duration of the war. Musical evenings were held in his mill. Dermot and Skem played the flute and Janet Gimson, a medical colleague of Dermot's and friend of Nancy's, sang. Janet is from a cultured and musical family, and is the great-niece of Ernest Gimson, who made beautiful arts and crafts furniture in the Cotswolds with the Barnsley brothers in the tradition of William Morris and Philip Webb. (She was later to marry Ken Roscoe, see p. 60)

In the early years of the war, Jas Pirie and David Green, the friends from undergraduate days, would drive out to Watford in their Lagonda cars to visit Skem and Nancy and gain some respite from the bombing.

Frank and Pamela Freeman, who had been witnesses at Skem's wedding,

Willows, Leaside, from The Wood Engravings of Mary Skempton.

were close friends from the Watford days. Frank was an aspiring but unsuccessful artist, always in debt. He had known Glossop when they were both undergraduates at Imperial College, and Silas got him a job helping at the Chingford laboratory doing tests. According to Pamela, he was a much better scientist than an artist. His paintings are, on the whole, gloomy and overworked oils, dark landscapes, which Pamela loyally had hanging in her house on the Isle of Wight. Pamela painted beautiful and detailed botanical watercolours. Frank was, however, a brilliant photographer. Before the war, he and Pamela lived in Chiswick in a circle of artists and intellectuals, which Skem later joined. This consisted of Julian and Ursula Trevelyan, Victor Passmore, Michael and Tanya Wickham, A.P. Herbert, and Tom Harrison of Mass Observation. Julian was a magnet, giving lots of parties in the large studio at Durham Wharf on the Thames at Hammersmith, where he and Ursula lived and worked.

Ursula describes finding the Wharf. On the day they saw it for the first time, peeping through a hole in the garage, they saw a naked man sunning himself. They thought, if he could do that, it must be secluded and quiet. They made a garden from what was a kind of engineering yard, with a raised area from which you could see the river. The trouble for Ursula was that Julian was very gregarious and a garden by the Thames was a mecca in summer. Julian painted energetically for short periods, joyfully abandoning the easel when people arrived, but the processes involved in making pots needs concentration over time. Ursula found herself making endless coffee and producing meals for the friends who were always dropping in. She would have liked more time for work. She thinks now she should have had a flagpole, like the Queen, and made it clear that, when the flag was up they were happy to have visitors but, when it was down, they were not to be disturbed.

Another difficulty for Ursula was that she endured years of miserable miscarriages. Julian was found to be responsible and agreed to take injections.

Ursula (Trevelyan) Mommens. Photograph by John Rastall.

When their son, Philip, was born, he was already a camouflage officer in the army, and took no part in rearing him.

Ursula says that she didn't know Nancy at all well, as (possibly significantly) Skem came to Durham Wharf alone. Ursula is a great-granddaughter of Charles Darwin and was brought up in Downe in Kent. (It was her younger brother, Robin Darwin, later head of the Royal College of Art, who introduced her to Julian – they had been at Cambridge together.) Although she denies it, Ursula must have been a very beautiful woman. Even now, in her 90s, she has high cheekbones and sparkling eyes and is an aristocratic and lively personality. I can well understand how Skem found her attractive.

Frank Freeman took wonderful photographs of this group. One particularly striking one is of Tanya Wickham, who was a spectacularly beautiful woman, half Russian and half Belgian, wearing a lace mantilla on her blonde hair, and nude above the waist. His photo of Victor Passmore gives him the cherubic dishevelled air that Dylan Thomas had in photos taken at about that same time. Passmore was much the best artist in this circle, in Skem's view, 'especially in his early more figurative phase'. Michael Wickham took a photo of Frank looking as handsome as Nureyev.

Michael Wickham's photograph of Frank Freeman.

The Freemans subsequently lived in a series of low-rent places, including a flat in the then near-derelict Trafalgar Tavern at Greenwich. For a time in 1941–42, while Frank worked in the Chingford lab, they lived at Cassio Road with Skem and Nancy. Pamela and Nancy shared the domestic chores and took care of me and the Freemans' eldest child, Sally. Frank had a razor tongue, and was always rowing with everyone. He and Pamela had a volatile but mutually dependent relationship. Pamela remembers a fight at the breakfast table at Cassio Road. Frank stormed out of the room and Pamela turned to Skem with a look, asking for support. He told her that she gave as good as she got. I can't imagine that Skem would have much time for these histrionics. Silas had an affectionate but despairing relationship with Frank, which is summed up in a rhyme he made up about him:

He stands outside 'The Good Intent',
His shirt is torn, his trousers rent,
His hair uncut, his bed unmade,
He dines on bread and marmalade.

(The Good Intent was a pub on the King's Road, Chelsea.)

After the Freemans moved out of Cassio Road to the next in their stream of temporary but artistic houses with low rents, their contact with Skem diminished and Frank had a last, fatal argument with Silas. Despite his other shortcomings, Frank was an active and involved father to his three children. He later lost an eye through neglected glaucoma and died in 1988.

Two or three architectural students were drafted into BRS as assistants during the war years. One of these was Ruth Pocock, a bright but rather awkward independent-minded girl. She had a long face and wore glasses and a serious expression. Skem says that she did not do well in her architectural exams at the Bartlett School because she chose to answer only those questions which she thought worth answering. Ruth had a habit of turning up unannounced at Cassio Road, where Beatrice was a frequent visitor and baby-sitter. Nancy told a story about Ruth with much amusement. One day, Ruth, who had never met Beatrice before, appeared up the stairs, encountered Beatrice and, out of the blue, enquired politely of her, 'Good afternoon. Do you like Scarlatti?' This to a woman whose idea of music was a brass band!

Frank and Dorothy Drake were neighbours at Watford and became close friends. They later moved to Bridgewater in Somerset, where Skem and Nancy visited them, and Nancy painted an oil of the local brickworks.

Nancy was eager to learn the piano in order to be able to accompany Skem on the flute. A Cassio Road neighbour was a pianist, Margot Osler, who taught music at Watford Grammar School and gave recitals locally. Nancy took piano lessons with her for a time. I remember her as an unhappy woman. When she had some kind of breakdown, Nancy very much took her under her wing.

Skem and Nancy's holidays at Cadgwith and contact with John Tunnard continued. The Guggenheim Jeune exhibition had found Tunnard at his most exuberant. Peggy Guggenheim describes the occasion in her autobiography: 'A woman came into the gallery and asked, "Who is John Tunnard?" Turning three somersaults, Tunnard... landed at this lady's feet, saying "I am John Tunnard."' He was known for his loud checked suits and garish ties. At the time of Skem's marriage, Tunnard, a conscientious objector, was acting as an auxiliary wartime coastguard at Cadgwith and breeding Blue Beveren rabbits as an additional source of food and income.

He wrote to Skem, 'I am writing this scribble to you from the look-out hut on the cliffs. It is blowing 70–80 mph from the east straight into the face of the hut. It is 6.10 a.m. and the sun is coming up red through an easterly gale haze.' He went on to describe his small holding, '11 laying pullets, 8 apple trees, 12 goosegogs, 18 loganberries, all new, and 18 square feet of horse manure in the cellar under the sitting room for mushrooms – with luck!'

Tunnard painting based on Skem's drawing

In the autumn of 1943, a slip occurred in the west bank of the Eau Brink Cut, River Ouse, to the south of King's Lynn, Norfolk. The geology was one of postglacial Fen clays and silts. In a famous paper in the Journal of the Institution of Civil Engineers published in 1945, Skem analysed the stability of the bank (a 'pioneer effort'). Meanwhile, Tunnard was still finding plenty of time for painting among his coastguard duties. On a visit to Cadgwith in 1944, Skem spoke to him of the work on which he was then engaged – the design of a flood relief channel in the Fens (coming down from Denver and running along, parallel to the Ouse, to King's Lynn.) Tunnard was devoted to the fenlands, where he had spent his boyhood, and was very interested in the designs that are generated by science and modern technology. Skem sent him a drawing of the Fellenius analysis (1936) of embankment design and Tunnard later painted two pictures, their composition based on this drawing, which perfectly combine his interest in science and landscape (Glossop, 1977). Skem has hanging in his room at Imperial College the resultant beautiful oil, and he also owns a watercolour, (unknown to the art world), which is a study for Tunnard's painting *Phi=0 at Wiggenhall St. Peter*. In 1977 he lent the oil to the Tunnard exhibition at the Royal Academy. Tunnard's self-portrait at the National Portrait Gallery shows a bespectacled face, juxtaposed with a detailed depiction of a sawfly.

Julian Trevelyan vividly described Tunnard in an article in the *London Bulletin* (quoted in Peat and Whitton, 1997): 'Who is John Tunnard? Why, he's the man in the loud check coat. You must have seen him, the man who always wears a shocking-pink tie, and a face that's a mixture between a fox's and a giant panda's. The man who laughs like a jackal, so that you hear him two blocks away. Don't tell me you haven't heard of him. The man who is always talking about shipwrecks and pirates. They say he lives in Cornwall, where he

undermines the morals of the older fishermen: he turns them into jitterbugs…'
In his autobiography, *Indigo Days*, Trevelyan wrote, 'After lots of drinks in the
pub and a session at his cottage during which he put on his pink plastic bowler
hat and danced to the records of Cab Calloway and Fats Waller, he finally
stumbled off round the cliffs to his (coastguard) look-out'.

Ursula remembers that, when Tunnard got tired of rural life he would
come up to London and dance madly with the Trevelyans at a night-club
called the Boogie Woogie. One night, his shoelaces came untied, so 'he put his
foot up on a chair, and all of a sudden rabbit droppings hopped out of his
trouser turn-ups. You never knew what was going to happen with John!'

In 1947, John and Bob Tunnard moved to a cottage on Morvah Hill. It
faced west and there were the most fantastic sunsets. When Skem and Nancy
visited them there, they stayed with 'Farmer Green' in his welcoming but
cramped and primitive farmhouse nearby. Skem watched in fascination as,
once a week, Farmer Green removed the flat cap that had seemed welded to his
head, revealing a pallid and bald pate. Peering into a small mirror, he then
began to shave his face, which in marked contrast to his smooth white head,
was ruddy and weather-beaten.

These artists were also musicians. Tunnard was a jazz devotee and had
started, and played the drums in, a successful student dance band at the Royal
College of Art. Julian Trevelyan's tastes were more classical and he played the
oboe rather badly, Skem remembers, but very enthusiastically.

Meanwhile soil-mechanics work continued at BRS. In May 1944, Skem
was invited by W. Fisher-Cassie to give a course of four lectures and
accompanying tutorials at King's College, then part of Durham University
(now the University of Newcastle-upon-Tyne) on 'A New Aspect of
Engineering Geology – Soil Mechanics and its Practical Application'. In some
biographical notes which Skem later compiled for the Royal Society, he writes,
'This was the first course of lectures I had given. I thoroughly enjoyed the
occasion, and it seemed to be a great success.' It was these lectures that inspired
Penman among others to pursue a career in geotechnical engineering.

In 1944, Silas Glossop finally settled down, aged 44, and married Sheila, a
petite, elegant, and efficient woman, 15 years his junior. Joan Souter-Robinson,
Frank Freeman's first wife, painted her portrait. When asked what attracted her
to Silas, Sheila laughingly replied, 'When you meet a man with a flat in Sloane
Square, you don't turn him down!' Sheila is a passionate supporter of the idea
that a woman's role in life is as the supporter of a man, and she provided Silas
with a well-ordered civilized background and encouraged him to build a
relationship with his daughter Susan. As Sue grew up in the chaotic household
of Antonia White, she came to prefer the relative calm (and regular meals) of
Sheila's household. In 1950, Silas and Sheila's daughter, Emma, was born.

By 1945, while the United States was dropping atomic bombs on Hiroshima
and Nagasaki and the war was coming to an end, Skem was becoming restless

Announcement for Skem's lectures at the University of Durham.

University of Durham

KING'S COLLEGE, NEWCASTLE-UPON-TYNE

DEPARTMENT OF CIVIL ENGINEERING

A Short Course of Lectures and Tutorial Classes
to be given in the College by

A. W. SKEMPTON, M.Sc., A.M.Inst.C.E.

A NEW ASPECT
OF

ENGINEERING GEOLOGY

SOIL MECHANICS
AND ITS PRACTICAL APPLICATION

1. The Geology and Mechanical Properties of Soils
2. The Stability of Slopes
3. Earth Pressure
4. The Bearing Capacity and Settlement of Foundations

Lectures at 6.30 p.m.
on May 8th, 9th, 10th and 11th, 1944

Tutorials at 4.30 p.m.
on May 9th, 10th, 11th and 12th, 1944

at BRS. Penman would say that one reason for this was Cooling's hesitation about progressing and publishing a special report they were writing together. It was rumoured at BRS that this was planned as a book, an *Introduction to Soil Mechanics*. Skem now says it was as well this was never brought to fruition, as the subject was developing and changing so rapidly at that time.

Cambridge dilemma

Be that as it may, in 1945, John (later Lord) Baker, Professor of Engineering at Cambridge University, who had heard of Skem through Pippard, 'Asked me to call on him after one of my excursions to the Fens early in 1944 with a view to discussing the possibility of starting soil mechanics research and teaching. I was very excited by this suggestion.' Skem was duly elected to a Fellowship at Trinity Hall. He continues the story (in his Royal Society biographical notes):

> As I thought more about what was involved I became increasing worried. The snag lay in the fact that, as the only Fellow in Engineering, I would have to supervise the students in all engineering subjects in their first and second years. This meant, in those days, at least 12 hours a week supervising studies in structures, hydraulics, thermodynamics, and indeed everything except electricity which would be looked after by a Fellow from some other College. Moreover the one subject apart from soil mechanics which I really wanted to teach, namely engineering geology, was not in the curriculum; nor did Baker see any chance of it being included.

Penman tried to help Skem revise by cycling over to Cassio road with his hydraulics notes. Skem also consulted his confidante Ursula Trevelyan. He told her that the first night at Trinity Hall, he thought this was the life for him. Sparkling dinner conversation, fine wines, an elegant suite of rooms. On the second night, he noticed that there was a certain amount of repetition in the sparkling conversation, he noticed certain tensions among the dons around the table, the rooms seemed dark and gloomy. 'What happened on the third night?' asked Ursula. 'I fled,' Skem replied. In a recent taped conversation with Ursula he mused, 'There is a lot about Cambridge that is very seductive. Such a lovely place, but I came to realize that it would be even worse doing what I didn't want to do among such beautiful buildings. Above all, I would be very bad at trying to teach these young men subjects that I personally wasn't very interested in, and indeed wasn't very good at. It would have been fatal.'

Skem wrote a fuller explanation of his decision:

> Now I knew that my powers, such as they were, lay primarily in field work and in the interpretation of field observations. I thoroughly enjoyed experimental work, but not, in general, as an end in itself, and I could see that there were almost limitless possibilities for pursuing my subject along the lines so successfully developed at BRS. But this required close contact with consulting engineers and contractors; otherwise the opportunities for work of the right kind would simply not arise. Above all I realised, though it was a hard struggle fully to admit to myself the truth, that I could not manage to do all that I knew to be necessary in my own work if I had to spend so much time, and for me so much effort, in teaching subjects which were far removed from my interests – soil mechanics and geology.

As he had not been educated at Cambridge, the systems there were unfamiliar. He offered to teach soil mechanics to all students, no matter which College they belonged to, but this did not conform to the Cambridge way of doing things, and could not be agreed to. He abandoned the Fellowship and left Cambridge before the beginning of the academic year.

Nancy had known, before he could admit it to himself, that it was not the right thing for him. The university post was, in the event, taken by Janet Gimson's husband, Kenneth Roscoe, with a Fellowship at his old college, Emmanuel; and Bernard Neal came in to the Trinity Hall engineering Fellowship. Both of them were well suited to the Cambridge engineering school. Although Skem found Roscoe a single-minded and unbending personality, he admitted that he built up a fine soil mechanics research lab. 'But he did little field work and was, I believe, never involved in a practical engineering job.' 'The decision to abandon Cambridge, and Professor Baker, for whom I had great respect and with whom I remained on very friendly terms, was appallingly difficult. It is one which, though correct, I have never ceased to regret the necessity of having to make.'

Since their work together on Chingford, Skem had kept up a correspondence with Terzaghi. In a letter from BRS of August 23rd 1945, Skem writes, as a younger man to an older respected mentor,

Your paper on stress conditions for the failure of saturated concrete and rock is... a most interesting paper and one which, like so many of your publications, I shall read over and over again... It still further enlarges the scope of our studies in the quantitative aspects of engineering geology.

Skem goes on to discuss the frictional properties of clays, then reveals the extent to which he was beginning to bracket himself with Terzaghi as two pioneers in a new science.

With regard to the relation of soil mechanics and engineering geology, I was very delighted to hear your views. I am hoping to take some action on this point by giving a series of lectures on engineering geology, which includes a good deal of soil mechanics as well as aspects of soil erosion, silting of reservoirs, properties of rocks, geology of dam sites, and water supply... We have to work untiringly for the recognition of engineering geology as a major subject in civil engineering and for the acceptance of soil mechanics as one of the essential branches of this subject... It is a great encouragement to know that you are sympathetic to this point of view. (Letter 23.8.45)

In England until this time 'the establishment' at the Institution of Civil Engineers did not recognize soil mechanics as a separate discipline of the profession. However, in 1945, the now-famous series of four lectures was arranged, and published the following year under the title of 'The Principles and Application of Soil Mechanics'. The other lecturers were Silas Glossop (whose knowledge of soil mechanics had proceeded at such a pace as to qualify him for such public exposition of his ideas), Markwick from the Road Research Laboratory, and Cooling. They were paid 20 guineas each. Skem wrote in his biographical notes:

This was a wonderful opportunity and I remember the pleasure of being able to devote several months to the preparation of the text and illustrations of my lecture on "Earth Pressure and the Stability of Slopes."...The series was well attended, and created a good deal of interest throughout the profession. The lectures were published by the Civils in 1946. I think it is fair to say that they finally established the place of soil mechanics, so far as this country is concerned, as one of the basic disciplines in civil engineering, both from the scientific and practical points of view. It led to the acceptance by the Institution of Civil Engineering of soil mechanics as a respectable engineering subject.

Skem told me that this lecture marked for him his 'coming of age' as a professional engineer. He considered that, of all his contributions, this lecture and his Rankine lecture of 1964 probably had the widest influence. Harold Harding was sitting next to Bill Allen and, while Skem was speaking, he leaned over to him and whispered, 'Isn't it amazing – this is the first time I've heard philosophy at the Civils!'

Meanwhile, Skem had not actually resigned from BRS and simply continued working there. However, that same summer, Pippard phoned him from London to say that he was intending to introduce soil mechanics at Imperial College, that he would like Skem to take on the job and that he hoped there would be a readership in due course. Skem accepted immediately and enthusiastically.

Chapter 4

Imperial College: Move to London
1945–55

Personally as well as professionally, 1945 was a turning point for Skem. The Cambridge decision was made and the decision to move to London. On the national stage, VE (Victory in Europe) and VJ (Victory over Japan) Days were celebrated and Churchill's coalition was replaced by a Labour government elected under the slogan 'Let's Face the Future'. The Beveridge reforms introduced the welfare state, and there was a feeling abroad that 'A new Britain must arise like a phoenix from the ashes of the fire raids'. (Angus Calder, *The People's War*.) However, along with this enthusiasm and, in Skem's case, the excitement of a new job, there were anxieties. Post-war Britain was an austere place, rationing continued, and shortages of basics, such as bread and butter, got worse. Nancy was pregnant again and Skem had money worries, doubting that he and his family could live in London on the salary Pippard was offering – £120 per annum for the post of Special Lecturer.

Skem writes of Pippard's job offer:

> This was a very different proposition from the Cambridge affair. My tasks would be to teach soil mechanics and advance the subject to the limit of my powers. Moreover there was already in existence the excellent course in geology, given by Frank Blyth, which had been such an inspiration to me as a student eleven years earlier. Pippard agreed that as soon as I had got the lectures and a lab started, he would appoint a young lecturer as my assistant. I discussed the whole thing with the Director of BRS and he allowed me to go up to South Kensington one day a week during the academic year 1945–46 and more often on occasions.

To get things started, Pippard asked Skem if he would give a course of lectures to the third-year engineering students at Imperial College, beginning in October. Skem describes the situation as he found it:

> A small unused laboratory in what was then the top floor of the Goldsmiths' Building, on the corner of Exhibition Road and Prince Consort Road, was available, though filled with miscellaneous discarded apparatus of one sort

and another and in a state of disrepair after four years of neglect during the war. Also there was next door a room, well lit by a skylight, which would make quite a good office, big enough for two people. Pippard agreed to allocate this room and the lab for my use.

(Address on the 50th anniversary of the study of soil mechanics at Imperial College)

Terzaghi wrote to Skem, 'I wish you the best of luck at this turning point of your professional career,' and sent a photograph of himself for Skem's new office. This photo still looks over Skem's desk in his room at Imperial College.

Skem had got to know Janusz Kolbuszewski when he visited BRS for a short period in 1943. He had been at Lwow University before the war, then joined the Polish Army as a young officer, and had escaped to London in 1941. There, he joined the refugee Polish University College in South Kensington to continue training Polish students whose studies had been interrupted. 'He was keen on soil mechanics, and he had a breadth of interest in the arts and sciences which greatly appealed to me. With Pippard's consent, it was arranged that he should enter Imperial College as a research student in October 1945. The first task he and I undertook was to clean out the lab, wash down the walls, and hang some boards with suitable photographs and drawings. I took a week's leave from BRS to do this job with Janusz. I shall always be grateful to him for coming at a time when there were virtually no existing facilities for research, and for the heroic work he did in helping to start the lab.'

Meanwhile Skem had written out a complete course of lectures. This lecture notebook still exists and is compiled in amazing detail. With only a little expansion, it could have been published as a short textbook. He started giving the lectures to about 20 of the third-year students in October 1945 and continued giving one a week throughout that academic year, in one of the fine old Victorian lecture theatres in the Waterhouse Building.

Back in 1938, Mr Manton, Lecturer in Highways at Imperial College, had visited BRS with a view to including some soil mechanics in his course on road construction and had, before the war, built two dead-load shear boxes, two oedometers, and a liquid limit apparatus from drawings which were given to such visitors.

These pieces of equipment had never been used, so we brought them up from the storeroom to the new lab. We purchased a balance and a constant-temperature oven for water-content determinations, and made a constant-temperature bath for specific-gravity measurements. In addition, I had a Proctor cell and a constant-head permeameter made in the college workshop from drawings I brought from BRS. With this equipment, the students could carry out a few experiments in the afternoon, following the lectures. As for research we decided to start on an investigation of the factors controlling the density of packing of sands, and to devise tests to determine the limiting loose and dense porosities. This work could be done mostly with apparatus made up from glassware obtainable from the chemistry stores and sieves, with the addition of a good microscope, another of our early purchases. Janusz threw himself with great energy into this work. He worked as an unpaid assistant in the students' lab course. The work progressed very satisfactorily and led to his PhD, 'the first from my lab', as I proudly note in my journal under the date August 13th 1948. Cooling was the external examiner.

Skem continues his account:

The students carried out the following tests: liquid and plastic limits, specific gravity, compaction, permeability (including a demonstration of piping) and shear-box test on sand. No marks were awarded for course work in soil mechanics in the first two years, but a question or two appeared in the exam paper known as ' Civil Engineering 2'. This absence of marks in no way affected the keenness of the students.

His full-time appointment as Senior Lecturer was approved in May 1946, to operate from October that year at a greatly increased salary of £700 per annum. At about the same time, Alan Bishop (of the Metropolitan Water Board) was appointed Assistant Lecturer. The previous year, the two had discussed the possibility of Bishop joining Skem at Imperial College and he was delighted when the appointment was made. 'Our views on soil mechanics were very similar, we both valued fieldwork and, if anything, he was even more intent on the practical engineering applications. But, in some respects, our talents were complementary; I was more aware of and certainly knew far more about geology, while he was far the better designer of apparatus and a remarkably skilled experimentalist.'

Bishop, Skem told me, had the 'most rigorous mind', he was 'remorseless, fearless, relentless' in his logical pursuit of an argument to its conclusion. He and Skem went for long walks to the Round Pond in Kensington Gardens, talking about scientific problems. He was a stimulating companion for Skem. He was slim, calm, reserved, unassuming. 'I loved him,' says Skem simply.

He continues: 'I had been for a long time deeply interested in Hvorslev's work on the shear strength of clays and indeed had outlined a programme of research in this subject to Ken Roscoe in 1945 when he came to seek my advice after taking the Cambridge job. For various reasons, Roscoe's attention was diverted to other topics so, by 1946, it had become a high priority in my own list of research projects. I therefore asked Bishop if he would design a set of three motorized constant rate of strain 6 cm shear boxes. This he did between May and July and they were ready to be used soon after the beginning of term in October. He also designed a twelve-inch shear box for testing gravels. This was made by the MWB but delivered to Imperial College, where, in November 1946, we began research on the shear strength of sands and gravels, including material from the site of the MWB dam at Walton-on-Thames.'

As Pippard had hoped when he offered Skem the job, in March 1947, he was able to appoint Skem University Reader in Soil Mechanics, the first in the subject in this country, and the summit of Skem's undergraduate ambitions. Skem wrote excitedly to Terzaghi to tell him the news. In his letter of congratulation, Terz replied, 'So far, really competent teachers in this field are very rare, because few of them realize the limitations which nature has placed on a purely theoretical treatment of the subject. Considering your previous work in this field and the concepts which you have developed during the years of your association with the Building Research Station, I believe that you are exceptionally well qualified to develop in your students a rational attitude toward the complex problems of earthwork engineering.' (Letter 23.4.47.) To complete Skem's pleasure, Alan Bishop was promoted to a full lectureship in October. It was that year that Philippa Wynne-Edwards (daughter of Sir Robert) joined the group as laboratory assistant and general handy-person.

Also in October, five research students arrived to join Kolbuszewski. They were Miss Agnete Lund, who started work on graded filters, and A.J. Smallman and J.B. Miners, both of whom Skem had got to know in the lab as undergraduates. They worked on the 6 cm boxes and left after a year, the first people to gain the DIC (the Diploma of Membership of the Imperial College) in soil mechanics. The others were M.A.A. Hafiz from Egypt, who carried out a monumental piece of work on the twelve-inch box, which led to his PhD in 1950, and I.K. Nixon, who worked on model pile tests in sand. 'How we located them all into the small lab is difficult to imagine but Pippard had already agreed to allow us to move down a floor into what was then the (much larger) Highway lab. Once again a massive operation of moving, wall-washing etc began, this time with more helpers and, by the spring of 1947, we were established in this fine laboratory, about twice the size of the old one. A triaxial apparatus, the first of Bishop's superb series of such machines, was installed in June. By that time, we had also built a tank for model seepage experiments, an automatic compaction machine and a model pile test apparatus… By the summer of 1947, then, the lab was very well equipped by the standard of that time.'

Skem wrote an account of his thinking about the study of soil mechanics in a letter to Arthur Casagrande at Harvard in July 1947.

I have found that it is quite possible to give (students) practical problems, often based on an actual job which I have done, and that they greatly appreciate this type of drawing office work… Mathematical treatment is discouraged since, in the whole of the past ten years, during which, at the Building Research Station, I have been working on practical problems, there has hardly ever been a case where any treatment other than straightforward statics has been necessary. On the other hand, nearly all the problems have needed an intimate understanding of site conditions and soil properties.

This viewpoint has been a key tenet of Skem's throughout his career and it later brought him and Imperial College into rivalry with the departments at, notably, Cambridge and Manchester who advocated a more mathematical approach.

To continue the story of early students and staff of the department, after Kolbuszewski took his PhD, the first in soil mechanics from Imperial College, he returned to the Polish College as Director of Studies and went to Birmingham University in 1950 to start soil mechanics there. R.E. (Bob) Gibson, an Imperial College graduate, arrived as a research student in the summer of 1947 and started work on the shear strength of clays, and R.D. Northey from New Zealand came to work on the structural sensitivity of clays. Skem says: 'I clearly remember, when Gibson first visited the lab in April, being impressed by his quiet competence and an evident devotion to research, and I know that Alan Bishop felt the same. We accepted him with enthusiasm and, of course, as events have proved, he turned out to be quite the most outstanding of our research students.'

Gibson told me how laborious he had found doing a drained test on soil samples in the old shear box, turning the top half of the box by hand until his arm ached. It had to be turned smoothly at an unbelievably slow rate, for a period of hours. Fortunately for Gibson, all joined the queue to take their turn, Skem included. Bishop's improved equipment with the electric motor was evidently much appreciated.

Janusz Kolbuszewski. Photograph by Peter Earthy.

The soil mechanics lab at IC in 1946. 'Bob' Gibson sits at the desk while 'Kolb' works at the far end. Photograph by I.K.Nixon.

Gibson described Skem's way of working with research students with this lively story: Kolb was doing a long series of tests on sands, using a funnel device which rained the grains down into a two-foot-diameter bin at a fixed speed. The funnel's width could be adjusted, and it could move up at the same speed as the sand level raised. One day Skem came up to Kolb and asked, with seeming innocence, 'How goes it? What are your results?' Janusz squirmed. 'Kolby, you have two variables here,' continued Skem. 'You have the height of fall, and the weight that falls on a unit area. When you increase the height of fall, do you get the sand denser in the container than when you have a lower height of fall?' Out of the corner of his eye, Gibson saw Kolb shifting uncomfortably. 'Well,' he said, 'with a greater height of fall, they are denser.' 'Why do you think this might be?' asked Skem. 'The grains acquire a higher velocity, they dig in between the other grains.' 'Yes, that sounds plausible,' said Skem. Kolb shuffled some papers in a drawer. 'These are the results,' he showed Skem, who having scanned them, remarked mildly, 'But the effect seems to be the other way.' 'Ah, that's because the grains fall faster, and hit the surface and bounce,' improvised Kolb, with an edge of desperation. If he'd remembered his results, he'd have saved himself. Skem said nothing, just walked quietly away, leaving Kolb to draw his own conclusions. Skem had a way of making people feel slightly uncomfortable without accusing them of anything, Gibson suggested, but his teaching method is congruent with androgogical methods of adult learning, as we now understand them, whereby students are encouraged to discover things for themselves, rather than be taught in a pedagogical way. (The foremost exponent of these ideas is an American educationalist called, strangely enough, Kolb.)

Gibson was aware of the health hazards of the early lab. These fine grains of sand were inhaled by students, who developed problems with their nasal passages. Once this was realized, measures were taken to protect them.

Skem's lecturing style has been universally admired. He prepares lectures with great care, and delivers them with the utmost formality. Lecturing to the students was his top priority (perhaps an excuse to avoid some committee meeting). Unlike Bishop, who was always announcing several days before a lecture that he was developing a cold and would not be able to lecture on Thursday afternoon, if you went into Skem's room shortly before a lecture, he would be looking at his notes and preparing himself. He sets out his material logically, clearly, and with humour. He has a reflective, even hesitant manner, sometimes pacing up and down as if searching for the *mot juste*, perhaps to allow the less brilliant members of the audience to catch up. His illustrations are well chosen and he uses physical imagery to express complex scientific ideas. He tells me that he has never given conscious thought to lecture technique, delivering them as he writes and thinks. He deals bluntly with speakers from the

floor whose questions are designed to demonstrate how clever they are, and is not afraid to say, 'I don't know,' if he doesn't. He once said to a staff member, 'Don't ever pretend to knowledge you don't have, never bluff.'

Gibson gained his PhD in 1951 but Skem told him that, as there was no lectureship available, it would be best for him to gain some practical experience. So Gibson spent a few years at BRS under Cooling before

Not without humour.

Skem called him back to IC as a lecturer in 1956. Gibson was a good mathematician and Skem valued this skill, as mathematics was not one of his strong points. They complemented each other in this respect. He regarded Skem as a valued mentor. I met him in retirement in Sussex, rotund and cheerful and a perceptive observer of the interpersonal dynamics among the early IC staff group. He knew that Skem was in a relationship with Ursula Trevelyan, and that she was part of Skem's life at about the time when he joined the department, having come across some letters, all in a similar handwriting, when he was looking for something in a drawer in Skem's room. He didn't read them, but put two and two together.

The last of the first group of staff at IC was the bluff South African, David Henkel. He had first met Skem at BRS. 'After spending the war years dealing with valves and time-based magnetrons, it was a breath of fresh air to find someone so full of enthusiasm for the aim of understanding the engineering behaviour of soils,' he says. He recognised Skem's 'talent and enthusiasm for understanding geology and its influence on geomorphology and the detailed engineering behaviour or soils and rocks'. Skem describes the development of the postgraduate course:

> Pippard was a strong advocate of postgraduate courses: there were already three running in the department (structures, concrete and hydropower) and he was making arrangements for a fourth to begin, on public health engineering. Early in 1950 he asked Bishop and myself to consider starting a course in soil mechanics, and enquired what would be necessary to carry it out. We replied 'David Henkel'. David was accordingly appointed lecturer. In October 1950, the first intake of five postgraduates arrived.

Soil mechanics was now established at college, in a form that changed little during the next six years, except that postgraduate numbers increased to about ten a year. Bishop was promoted to a readership in 1953 and Skem became a professor in 1955. David Evans came that year as a full-time technician. He stayed in the department for 25 years.

One of the first group of postgraduates, and the only American, was Don Roberts, who, looking at the courses he was to take, found that one was called 'Relaxation Methods'. 'I thought this must have to do with the fine art of beer drinking at the local pubs. I was disappointed to discover that this was really a course in applied mathematics and involved the construction of ground-water flow nets. Relaxing it was not!' He also describes triaxial analysis of the engineering properties of brown sugar. Several sugar silos in the Caribbean had ruptured suddenly and it was suspected that the sugar was behaving as though it were a viscous fluid rather than a granular material. (A triaxial test consists of putting the material inside a rubber sleeve and crushing this cylinder of material while it is subjected to lateral pressures of varying intensities.) The tests on the sugar did not harm the texture or flavour of the sugar and the soils lab was very popular at a time when, in the postwar austerity, each person's weekly ration of sugar was only four ounces. 'Students and faculty members would show up with cups of tea waiting for samples of the sugar after the tests were complete.' Roberts and his wife held an end-of-year party, with a rare ham from the States, and champagne cider flowing freely. 'Everyone got a little tipsy, even Skem' but, if the subject turned to soil mechanics, 'Skem's mind would suddenly become as clear as a bell.'

Ever the realist, Skem's decision to take the Imperial College job had been influenced, not only by his enthusiasm for the prospect but also by the fact that he could just afford to do so, thanks to an intervention by Silas Glossop. His lecturer's salary was to be £700 per annum but, after the rent was paid, there would be little left over for other expenses. Before taking up his new post, Skem had consulted the former wife of Frank Freeman (witness at his wedding), Joan Souter-Robinson, about the cost of living in London. (She was another inhabitant of Oakley Street, where Skem and Silas had had the *pied-à-terre*). She was doubtful that the lecturer's salary would be adequate to support a family. Since 1938, Terzaghi had been retained by Mowlem's as a consultant but the fact that, by this time, Terz was living and working in America made this arrangement inconvenient for both parties. Silas dropped a word to Sir George Burt, director of Mowlem's, saying that, despite his youth, Skem was the right man to succeed Terzaghi as consultant for the firm. The outcome was that he was offered a £100 per annum retainer fee with Mowlem's. This would be in addition to any fees he earned by carrying out specific consultancy jobs for them. In this way, Skem retained a strong link with industry and also supplemented his university income. Within a short time, Skem was getting many consulting requests and fees from this were a considerable additional income throughout his career.

In November 1946 Skem was asked by the Institution of Civil Engineers to join a newly formed Soil Mechanics and Foundations Committee as the representative of the universities. Other members of the Committee included W.K. Wallace as chairman, A. Banister (ICE research officer), Cooling, Dr. Glanville, Silas Glossop and Dr. Herbert Chatley. They decided to compile a bibliography covering the period 1920–46, to review the teaching of soil mechanics in the universities and colleges, and to draw up a syllabus for engineering geology and soil mechanics for the Associate Membership examination. The bibliography was published in January 1950, and, following the eventual agreement by the Institution Council, the first paper on soil mechanics and engineering geology was set in April 1951, and for several years Frank Blyth and Skem were moderators for this examination.

The Little Boltons, South Kensington

During 1946–47 Skem and Nancy continued living at Cassio Road with me, just starting at school, and Katherine, a baby only a few months old. He travelled up to London four days a week on the Green Line bus but spent one and a half days a week (including Saturday mornings) at BRS, finishing off various bits of work.

He had, meanwhile, to find somewhere to live in London. The first plan was to share a house with their good friends from student days, Jas and Psyche Pirie, but the Piries soon decided instead to build a modern house in Holland Park. At BRS, Skem had made a friend of John Eastwick-Field of the architecture section. He and his wife, Elizabeth, unlike Skem and Nancy, had some capital funds at their disposal and the two couples decided to buy a house to share in order to pool resources. As Nancy was fully occupied in Watford with looking after the new baby, Katherine, it was Skem who did the house-hunting in the intervals in the work of setting up the lab. After seeing a variety of totally unsuitable places, he found a rather grand tall white stucco Victorian semi-detached house in Kensington, at 24, The Little Boltons, SW10. The area, now

24, The Little Boltons, SW 10
and the family on the steps.

one of the most sought-after and expensive in South Kensington, was, in 1946, littered with bomb sites. A space where the next-door house should have been sprouted rosebay willowherb from among the rubble. Prices were rock-bottom. It was the time when many of the population were having to find shelter in Government prefabs, pending the rebuilding of their shattered streets. John bought the freehold and Skem rented a part of the house on a lease from John under 'a gentlemen's agreement' for £250 per annum (an arrangement which was to have dire consequences ten years later, when the lease expired). John used his architectural skills to convert the house into two maisonettes. John and Elizabeth had two children: Nicholas, who was my age, and Jacqueline, who was Katherine's. (During a subsequent pregnancy, Elizabeth contracted German measles and her fourth child, Stephen, was tragically born blind and with a cleft palate.) The Eastwick-Fields were to occupy the ground floor and basement of the tall house, and the Skemptons the first and second floors. Our rooms were light, spacious, and high-ceilinged. The sitting room, and the bedroom above it that Katherine and I shared, looked westwards up Redcliffe Square. The kitchen at the back overlooked the garden below and got the morning sunshine. There was a coal-burning stove with little opening doors in the living room, which Skem used to sit beside in his wheel-back armchair. From Cassio Road came the Bechstein upright piano and a grey cupboard and sofa for the living room but, otherwise, Skem and Nancy had to furnish the entire flat on a shoestring budget. To help with the rent, Skem let the spare bedroom to Nicola Darwin, a great-granddaughter of Charles Darwin and sister of Ursula Trevelyan. She sang soprano in the Bach Choir and gave me piano lessons. (She later made a somewhat unsatisfactory marriage to a man everyone thought had little to say for himself, moved to the West Country, and died quite young.) Hugh Pite, who also lived in our part of the house, was an architect friend of the Eastwick-Fields and, technically, their lodger.

Towards the end of the time at The Little Boltons, when he moved out, his room became our dining room and Skem had to pay extra rent to Eastwick-Field for that room. Stone steps led up to the shared front door and by the stucco front wall were a pair of lime trees, which dropped their sticky deposits on any toys left out overnight. The back garden, with its gnarled may tree and, soon, a sandpit and swing, was to be shared between the two families. A neighbour two doors down was Phillip Powell of the famous architectural partnership, Powell & Mora.

By September 1947, Skem and Nancy and the children were established at The Little Boltons and Skem spent every day thereafter at college only going to BRS occasionally to keep in touch, thus ending exactly eleven years' association with BRS.

Following the family pattern, Skem found a good prep school for me close by, St David's in Elvaston Place, SW7, off Gloucester Road. He used to walk me to school en route to college. When Katherine was a toddler, she went to the kindergarten of the Lycée Française, Le Jardin des Enfants, which was in a large house at the junction of The Boltons and Gilston Road, and she followed me to St David's. Henri Cartier-Bresson took a famous photograph of 1950s London schoolchildren, which is clearly a crocodile of St David's children lining up in Elvaston Place, supervised by the son of the headmaster, John Durnford, elegant in a suit with a rolled umbrella.

The years 1947 and 1948 were glory days for Skem. His department was established and he had exceptional and congenial colleagues around him. He, Bishop, and Henkel formed a powerful trio with complementary strengths. The profession was becoming recognized across the world, with Imperial College as a centre of excellence. He had the full support and gratifying recognition from his mentors, Terzaghi and Pippard, whose faith in him was bearing fruit. His work was utterly absorbing and breaking new ground. He also had a beautiful, creative, and very capable wife, two healthy daughters, and a pleasant place to live.

Nancy

I do not know whether Nancy knew of Skem's relationship with Ursula. It was certainly never discussed in my hearing. It was also of fairly short duration. Ursula soon left Julian and later married a Belgian potter called Mommens, and they set up a pottery together near Newhaven, East Sussex. Skem remembers Ursula and Mommens visiting at The Little Boltons. Meanwhile Nancy herself had, I think, a soft spot for Jas Pirie, her undergraduate friend.

By 1947, Nancy had some major decisions to make. She was finding that the inspiration for her painting was difficult to maintain while caring for two small children and a husband who was spending huge amounts of energy at his work and away from home. She needed (and wanted) to give sustained attention to her family, and not have to break off to boil potatoes, change nappies, organize the laundry. The countryside, which she loved and which had inspired her painting and wood engraving, seemed far away from her South Kensington existence. She also perceived, in the ten years since leaving the RCA, a lack of success as a wood engraver. As Skem writes in his foreword to a volume of Nancy's wood engravings, which he published privately towards the end of her life:

> A couple of dozen prints sold privately or through exhibitions and about half that number of minor commissions seemed scarcely to be the basis for a future career. This assessment may have been too modest. Nevertheless, she decided to change direction and become a bookbinder. Engraving from now on was to be simply for the pleasure it gave to her, and to our friends.

She started sending out wood engravings each year as Christmas cards, which are still much treasured by their recipients, but they tended to be sweet studies of small animals, which, I think lack the inspiration of her earlier Cornish village landscapes.

Nancy decided to build on the bookbinding training she had begun at the Royal College of Art under William Matthews and returned to work with him one day a week at the Central School of Arts and Crafts in Holborn. In order to make this possible, Beatrice came up to London from Watford every Thursday to take care of my sister and me. Nancy did this for five years and kept in touch with Matthews until 1957.

From her workroom at home some forty commissions were completed by the end of 1951. They included the first set of *Signature* volumes: two of these bindings are among the finest she has ever done. (*Signature* was a journal describing itself as, 'A Quadrimestrial of Typography and Graphic Arts'. It is no longer in publication.)

Skem continues:

> One Thursday a year or so later she brought Matthews back for tea. A man of few words, he smiled as I showed him out of the front door and said, 'She'll do'; this I took as high praise from a master craftsman.

Despite the fact that they had moved away from Watford, the influence of Beatrice on the Skempton household was considerable. The family spent many weekends at Orchard Drive, enjoying Beatrice's cooking, sitting in the garden and walking by the Grand Union Canal and around the Hertfordshire lanes. Nancy told me that, in the early years of their marriage, while she was baby-sitting for us, Beatrice would pair the socks in Skem's drawer, to Nancy's, no doubt unexpressed, fury. The bond pulling a determined widow and her only son together could not be easily relaxed. Much as she loved and admired Beatrice, Nancy was very much aware of her continuing power over her handsome husband.

Judith and Katherine, by Frank Freeman.

Nancy was a devoted mother to my sister and me. She provided us with the unconditional love that Skem could not give, busy as he was. He only took notice of our academic progress. She was creative in providing us with special things, at a time when toys and clothes were hard to come by. She made a dolls' house for us, with real glass windows, a back door, and a roof 'tiled' with corrugated cardboard painted deep red. She also sewed a nursery frieze, a collage of a street scene made with bits of cloth and fabric, complete with milkman delivering milk and a town hall topped with a cupola.

Much later, as a young supporter of the women's movement in the 1970s and 80s, I used to express anger with Nancy for the fact that she had so subordinated her own talents to those of Skem. She invariably made his wishes and needs her priority. She saw her main role in life as supporting him so that he could concentrate on his work, even to the extent of learning a considerable amount about soil mechanics herself. To counter my anger, she told me quite firmly (and rightly) that she had made her own choice about how she wanted to lead her married life and that I need not question her decision. Undemonstrative as he was, Skem nevertheless valued and respected her in the most fundamental way, not just for her devotion to him and his needs, but also for her sound common-sense approach to life and, above all, her artistic creativity.

Rotterdam conference

The famous 1948 Rotterdam conference was a revelatory experience for Skem and all who were there. It was the second International Conference of Soil Mechanics and Foundation Engineering. The first had been held at Harvard in 1936, with Terzaghi as its star, with the intention that one would be held every four years, but the war had intervened. Leonard Cooling had been the only British engineer at the Harvard conference among 200 participants. During the war, there had been little communication and even less publishing. Rotterdam was an occasion for meeting each other for the first time and finding out what everyone had been doing. Terzaghi was the president of this conference, with Skem as his deputy. Nearly 600 delegates attended and there were no less than twelve papers from Imperial College. The proceedings fill all of seven volumes.

For Skem, it was the first of many important conferences at which his work was presented. Skem, together with other members of the ICE Soil Mechanics Committee (by now called the British National Committee), had been instrumental in assembling papers from colleagues. In his biographical notes he writes,

> The response was amazing, and at a meeting in September 1947, having sifted out the submissions, we accepted 76. The authors were then told to prepare their contributions and submit them to the British National Committee not later than 1 December 1947... Again the response was remarkable and a large batch was sent to Holland.

The high table at the 2nd ISSMFE Conference in Rotterdam, 1948 with Skem on Terzaghi's left.

Unlike future conferences, on this occasion Nancy did not accompany him, because Katherine was still a small baby. (Ursula Trevelyan spent a weekend in Rotterdam with Skem, very discreetly, with every effort made not to upset Nancy or Julian. I certainly don't remember any repercussions from this secret time.) The excitement of meeting like-minded colleagues from Europe and exploring a new subject is indicated by a note Skem made in his journal: Shear strength: people interested – Haefeli, Kjellman, Geuze, De Beer, Taylor, Hansen, Bjerrum.

The papers were actually published before the conference and, in the conference hall, discussion took place on the issues raised. There can have been no extravagance, as the British Government, as part of post-war austerity measures, had limited daily expenditure to £8 for those travelling abroad on business.

The American civil engineer, Ralph Peck, explains the excitement of the occasion: 'The burning question in soil mechanics was why the shear strength of soft saturated clays in the field seemed to be predicted best by laboratory tests on samples that were not first allowed to consolidate under the stress conditions to which they had been subjected in the field. Instead of using triaxial tests that presumably subjected the samples to the initial state of stress in the ground and then applying the stresses expected as a result of construction, the best agreement with observations of failures in the field was obtained simply by taking the shear strength as half the unconfined compressive strength of "undisturbed" samples from the field. This was very convenient, because both the testing and the stability computations were much simplified, but it defied logic. Skem addressed this subject in five papers or discussions in the Rotterdam conference, and clearly made the point of view (then called the phi = 0 concept) respectable.' (Personal communication) The most notable paper was 'A study of the geotechnical properties of some post-glacial clays'. Professor John Burland of Imperial College sums up Skem's contribution thus: 'He demonstrated for the first time that the undrained strength in compression is proportional to effective overburden pressure in normally consolidated clays, and that the factor of proportionality is about 0.3 – a figure which is widely accepted today.'

These papers were the culmination of Skem's research from BRS and early Imperial College days, and they were all based on research generated through actual consulting work on 'jobs', though validated by theoretical work.

Skem met T.K. Huizinga and his senior member of staff, E.C.W.A. Geuze, for the first time at Rotterdam. According to Silas Glossop, their laboratory in Delft was the largest in the world at the time and they had developed a system of soil testing based on the work of Professor Buisman, which differed in many respects from that of Terzaghi's school at Harvard. (British Geotechnical Society Twenty-fifth Anniversary Report. *Géotechnique* 25. No. 4 1975) In his album Skem has a photo of Terzaghi and Huizinga examining some triaxial testing equipment in the Delft lab. A party from the conference visited polder

Terzaghi and Huizinga in the Delft lab.

Skem on Zuider Zee dyke.

drainage schemes and another photo shows Skem looking windswept in tweed jacket on the Zuider Zee dam.

Beginnings of Géotechnique

Developments in soil mechanics were so rapid in the immediate post-war years that the need began to be felt for a specialist journal, and preferably one of international scope. Glossop, Golder and Skem led the small group of friends calling themselves the Geotechnical Society, who met this need by initiating the publication of the journal *Géotechnique*. I cannot do better to describe the genesis of *Géotechnique* than by quoting Silas Glossop's account in the British Geotechnical Society Twenty-fifth Anniversary Report (*Géotechnique* 25, No. 4 1975): 'During the War we had been isolated from workers in all other countries except the United States. So, at the first opportunity, in 1946, I asked permission from Mowlem's directors for Golder and myself to visit the principal laboratories in Western Europe.' At Delft, they met with Geuze. 'Now all those who have met him will know that Professor Geuze is an outstanding person both intellectually and socially, so after a most interesting day we asked him to dine with us in Rotterdam. After dinner, at which we talked of nothing but soil mechanics, we went on to a nightclub. There, undistracted by the floor-show, we continued to talk soil mechanics. Certain it is that on that occasion, the suggestion was made of starting a journal… We then had several decisions to make. Who was to publish it? Who was to edit it? And how was it to be financed?' Another visit on the trip was to Lausanne, where it was the French-speaking geotechnical engineer, J.P. Daxelhofer, who suggested the title, *Géotechnique*.

'We approached our friends, Cooling, Skempton, and Ward and, between us, we decided to set up a geotechnical society, primarily to publish the journal, but also to further the cause of soil mechanics in any way that seemed possible.'

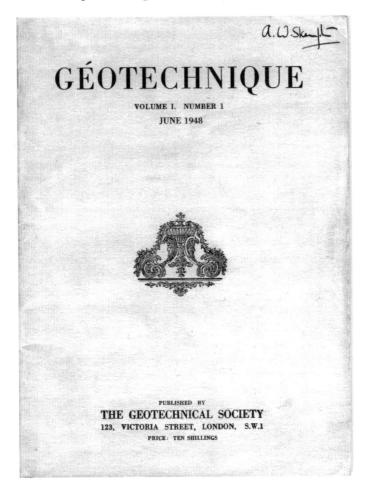

The first edition of Géotechnique.

Their first step was to send out a prospectus or circular letter and the response was overwhelming. Authors were approached and articles promised and the journal began to take shape.

Silas continues the story. 'I called on my own bank, the South Kensington branch of Williams Deacon's, where I had opened an account for the Geotechnical Society, and spoke with such enthusiasm on the future of our journal that the manager, Mr Esmond Barclay, a most charming man, allowed the Society an overdraft facility sufficient to ensure that volume one got through the press, thus gaining for himself a place in engineering history.'

Silas commissioned from Nancy the layout of the front cover, which incorporates the famous printer's ornament (a colophon) from Coulomb's paper of 1773 on earth pressures, and her format continues almost unaltered today.

Glossop and Golder were 'temporary editors' and they defined the aims of the journal in the editorial to the first issue which appeared in June 1948: 'To promote international collaboration between workers in Soil Mechanics and related sciences; to publish papers on specialized aspects of these subjects, such as might not be acceptable to the Proceedings of Institutions and Societies devoted to Civil Engineering; to encourage the pursuit of engineering geology; and to make the results of research available to the practising civil engineer.' All these four objects it has achieved. The first issue contained six papers, two in French, and this international focus continued throughout the early editions.

Glossop and Golder went on to produce four issues in 1948 and 1949. They carried most of the burden of the production of the journal in those early days and Glossop remained on the editorial board for 20 years. In 1949, they persuaded the Institution of Civil Engineers to take on the publication of the journal and an advisory panel was set up with Cooling, Skempton, and Guthlac Wilson joining Glossop and Golder as members – the Geotechnical Society in another guise.

Skem's work on the harbour at Gosport and at the alluvial river plain of the River Forth at Grangemouth formed the basis of the paper now famous among soil-mechanics engineers, 'The Geotechnical Properties of Some Postglacial Clays' which was published in this first edition of *Géotechnique*. Skem describes this paper (in his biographical notes for the Royal Society):

> It dealt with the changes in water content and strength for normally consolidated clays with depth, a topic on which I was engaged in friendly but strong controversy with Terzaghi at that time. His ideas were, surprisingly, rather eccentric on this subject, but I managed to collect and analyse sufficient field evidence to make a convincing case for what is now taken to be the orthodox view (which I elaborated in a massive paper to the Geological Society in 1969).

Doubts had been raised as to whether there really was a gain in strength with depth in soft clay, as there seemed to be evidence against this. In fact Skem found that the results which seemed to indicate that there was not a gain in strength were due to disturbance caused by the sampling process itself. Skem sums up the achievement of *Géotechnique* over the subsequent years, 'There can be no doubt that *Géotechnique* has been (and of course is still) a factor of the utmost importance in the soil mechanics world. It provides a journal of impeccable standard, both from the scientific and production points of view, for papers on soil mechanics, engineering geology and rock mechanics, which is read by everyone interested in these subjects. I am proud to have been associated with it so closely and for so long.' (Biographical notes for the Royal Society.) In fact by the time Skem retired from the advisory panel in 1972, he had served on it for 23 years, the last two years of which he was the editor.

Gibson, then a young research student aged 23 was, meanwhile, bored with endless tests in the lab and keen to publish his first paper in *Géotechnique*. He had worked overtime in some secrecy to prepare it and eventually plucked up the courage to discuss it with Skem, who soon discovered through Golder that the Danish Professor Hansen, a much more eminent scientist, a giant of six foot six with a rectangular Nordic face, was writing a paper on the same subject. Skem knew this would upset Gibson, so he wrote to Hansen, enclosing a copy of Gibson's paper, suggesting that perhaps the two papers could be united. Hansen readily agreed and Skem wrote the Introduction to the joint paper, which duly appeared in the first volume. Had Skem not intervened in this way, Gibson's paper would not have been published at all. In this way, Skem fostered the work of his protégés and also encouraged international collaboration in the geotechnical field.

Social events with Terzaghi and the Glossops

Terzaghi's personal charisma, as well as the strong influence he had on his young colleagues, made him the centre of attention. Everyone who knew him at that time has stories to tell about him.

Hugh Golder of BRS relates a tale about an evening of *rijstafel* at the home of Professor and Mrs. Geuze during the Rotterdam conference, where Skem, Glossop, and Terzaghi were among the guests. Geuze knew that Terzaghi only drank cognac. As he had plenty of Dutch gin but only one bottle of cognac, he asked his guests if they would please ask for the latter. He approached Terzaghi, gin and cognac bottles in hand. 'When I'm in Holland I always drink Dutch gin,' said Terzaghi and, taking the almost-full gin bottle, poured himself a glassful, and placed the bottle on the floor by his chair. The *rijstafel* was excellent, the evening stimulating, Terzaghi telling fascinating stories. At about 2 a.m., he picked up his bottle of gin, examined it, and found it empty. 'Now we go home,' he said, and they went.

As well as his personal magnetism, Terzaghi had prodigious energy. Skem remembers a particular occasion during the Rotterdam conference when a group stayed up talking until the early hours. At about 2 a.m., Skem, exhausted, had to go to his room for a lie-down. He emerged at 4 a.m. to find Terzaghi *still* in full flow. Terzaghi and Skem travelled back together from Rotterdam on June 29th, deeply into intense discussions on the cross-Channel ferry about Zeeweart's data on Mexico City clay. Skem was simultaneously engaged in studying slopes in a river valley at Peterlee, County Durham. The river had eroded the boulder clay of the valley over thousands of years and Skem had worked out for the first time the ultimate stable slope in boulder clay from field observations. At Terzaghi's suggestion, they chartered a small plane at

Bovingdon Airport and flew up to visit the Peterlee site. On the way back, Terzaghi insisted that the pilot circle several times around Durham Cathedral, which, together with the headwind they encountered on their way south, led to a nerve-wracking flight for Skem, who was painfully aware that the fuel gauge was reading near zero. Terzaghi, busily smoking large cigars, was evidently oblivious to the danger.

Silas Glossop had not encountered Terzaghi at Chingford and it was during Terzaghi's first visit to Europe after the war that Silas and Sheila held a dinner in his honour at their flat at the top of 6, Sloane Square. Silas recalls this event in a *Géotechnique* tribute to Terzaghi written after his death. 'The only other guests were the Skemptons. I knew that Dr Terzaghi enjoyed good food and good wine so, in the face of post-war rationing, I searched the market so that my wife could give a dinner party suitable for an appreciative guest.' Skem remembers that, astonishingly, Sheila served a whole salmon. Silas continues, 'The war was just over and Skempton and I, both of whom were deeply interested in soil mechanics, had had few opportunities of discussing the then-outstanding problems with research workers in other countries. There were two such questions of the first interest – Terzaghi's own theory of consolidation, and Horslev's work on the shear strength of clay. Our discussion with him went on through dinner and lasted until four o'clock the next morning.

At first light we went down the 97 steps from my flat to see Dr Terzaghi back to his hotel in South Kensington. My car was a pre-war model with no self-starter and he was amused to see me cranking the handle for, in the United States, hand-started cars were already a generation out of date. Several minutes' cranking would not start the engine, so I set off to find a taxi; and they were rare.

'In those days there was a coffee stall in Sloane Square and, as I walked into the Square, a dilapidated motor lorry pulled up alongside it to tow it home. I made the driver an offer to take us to South Kensington and he accepted. So the night ended with Nancy, Skem, and Terzaghi sitting with our feet dangling over the end of a broken-down lorry, which, firing on three cylinders, rattled off to his hotel.

'When we said goodbye to him on his own doorstep at 5 a.m. he was in high good humour. He had obviously enjoyed himself. So had we, and we had learnt much from him.' (*Géotechnique*, 14, 1964). How the Skemptons returned to The Little Boltons is not recorded.

Silas also relates the tale of the visit to Avebury. Terzaghi was passing through London on his way to India in October 1946. He missed his connecting aeroplane and so had to spend the weekend in England. Silas and Sheila Glossop and Skem packed into a large pre-war Wolseley motorcar lent by Sir George Burt of Mowlem's for a sightseeing trip. The resourceful Sheila packed a hamper full of wine and gourmet picnic dishes. Silas writes: 'At Winchester we visited the cathedral to see the great nave, William Wynford's masterpiece and, as engineers, to trace the enormous settlements of its foundation. Terzaghi enjoyed the story of how Sir Francis Fox had underpinned it with the help of his foreman diver, Bill Walker, to whom a monument has just been placed in the cathedral. From Winchester we went on to Dorchester to see the Museum and the Iron Age fort of Maiden Castle, which Vespasian's army took by storm nineteen hundred years ago. Then we walked along the great ramparts and, of course, estimated the man-hour content they represented, and speculated on the nature of the barbaric society which had built them. He was wonderful at this game and could always find two hypotheses to anyone else's one.' They proceeded to Stonehenge, and on to Avebury to see the gigantic and mysterious earthwork, Silbury Hill. Sheila took a photo of the three giants of soil mechanics picnicking and looking convivial leaning on the Neolithic embankment which surrounds the stone circle.

Congress of Applied Mechanics

In those early years, Skem was so busy on consulting work that he did not have time to write up each job, and concentrated on publishing those that presented issues that particularly interested him. His journal for 1946–48 records the intense packed summer of 1948, when every other day lists major events, and notes his thoughts and writing on a whole range of subjects. He was preparing

Glossop, Terzaghi and Skem at Avebury. Photograph by Sheila Glossop.

a paper on the effective stresses in saturated clay for the Seventh International Congress of Applied Mechanics, not the happiest of events for him. After he had delivered the paper, Professor Rodney Hill of Cambridge, an expert in the theory of plasticity and a brilliant mathematician, spoke from the floor demonstrating that the theory was thermodynamically impossible. After that experience, Skem rarely attempted any further applied mechanics.

Skem's continuing dependence on the advice and support of Terzaghi is illustrated by an exchange of correspondence between him and Karl's wife, Ruth, in January 1952. Skem had sent Karl his paper on the sensitivity of clays but Ruth wrote back that Karl was at that time away in British Columbia. Skem replied,

> If your husband has any terrible hammer-blow criticism, please ask him to deliver before and not after the paper is published. In writing every sentence, I had him looking down from his photograph on my wall, and he was constantly in my mind. But this is no substitute for his actual criticism and encouragement, which I value above all else in my work. (Letter 29.1.52)

Terzaghi's reply did not come until July, saying the paper 'is a landmark in our knowledge of the sensitivity of clays and it provides us with a coordinate system into which we can fit the results of our future observations'. But this praise is mixed with some ferocious criticism of Skem's paper on vane tests at Grangemouth, accusing him of 'daring over-simplification' in a diagram.

The Shear Strength meeting in London in June 1950, was, after Rotterdam, a local affair. The English crowd had decided to arrange it. European friends were invited, including Huizinga and Geuze, and Kjellman. Skem consolidated his close friendship with the tall blond Danish engineer Laurits Bjerrum, whom he had first met at the Rotterdam conference. He wrote of him:

> I was immediately impressed by his vivacity, by his knowledge and intellectual discipline and, above all, by the combined charm and firmness of his character... From the time of the shear strength conference until he died, he and his work were never far from my mind. I was frequently seeking an opportunity to discuss new ideas and new discoveries with him and looked forward with the keenest anticipation to hear of his latest research... among the many interests which to our delight we found we shared was a devotion to the beautiful work of Juul Hvorslev.

A dinner was held at a restaurant called Verreys and in Skem's album is a page from that evening covered with mathematical formulae and the signatures of the Imperial College crew, Alan Bishop, David Henkel, and Hugh Golder, with, among others Guthlac Wilson, Kenneth Roscoe, and Laurits Bjerrum. Harold Harding has added to his signature, '*Donc dieu existe*'. On the last day of the conference Nancy organized a party in the garden of The Little Boltons. Skem's album contains a photo, showing Bjerrum looking handsome in his shirt sleeves, talking to Glossop's wife, Sheila, and Golder's wife, Molly. That night Skem and Bjerrum stayed up until the early hours discovering very many shared interests.

> We shared a conviction that field work played an absolutely essential part in soil mechanics, that geology should always be considered, that the history of our subject should not be neglected, that Terzaghi was a genius of the highest order, and that we enjoyed every moment working in soil mechanics,

wrote Skem later. (The paper Skem and Bishop wrote for this meeting, 'The Shear Strength of Soils,' was published in the second issue of *Géotechnique*, in 1950.)

While Britain pulled itself together, started to overcome the post-war privations, and flocked to the life-asserting 1951 Festival of Britain on the South Bank of the Thames, the pattern of Skem's work was of intense research, which he then wrote up for a series of almost annual conferences of one kind or another. The work that Skem had done on foundations at Kippen back in 1940 (and which he had used for his AMICE) formed the basis of his paper, 'The Bearing Capacity of Clays' for the 1951 Building Research Conference in London. This paper was regarded by Golder as a 'little classic'. A shear failure had occurred in soft clay below a small, almost square building, a rare occurrence providing an opportunity to assess the then-disputed relationship between the bearing capacities of square and strip footings. (Golder, British Geotechnical Society Twenty-fifth Anniversary Report, p. 655) This paper is still used for the assessment of the stability of foundations in clay soils.

Usk, Daer, and Chew Stoke dams

In the early 1950s there were three significant embankment dam jobs for Skem. Binnie Deacon and Gurley was the consulting engineering firm involved in both the Usk and Daer dams, both constructed using boulder clay fills. BRS was asked to install piezometers into both dams to measure the pore

Bjerrum and Molly Golder at garden party at the Little Boltons.

pressures. Skem and Bishop analysed the findings and, at Usk recommended that 'drainage blankets' be introduced before each season's fill, thus restricting the future build-up of excess pore pressures. Penman writes, 'This success caused many future dams to be fitted with drainage layers, to such an extent that it almost became a fashion.' (*Dams & Reservoirs*, Feb 2002).

Skem was consulted before construction of the Chew Valley dam near Bristol by the consultants, T. & C. Hawksley, and recommended the installation of a system of 500 sand drains calculated to reduce the build-up of pore pressures within the clay. He installed Casagrande-designed piezometers placed in boreholes made through the fill into the foundation at several positions, which showed a very satisfactory performance by the sand drains. Skem considered this to be the first example where sand drains had been used as an essential part of the design, as distinct from their use as a remedial measure. He and Nancy were present at the inauguration of the dam by the Queen in 1956, and it is still retaining the Chew Valley Lake that stores water to supply Bristol. Skem and Bishop wrote a paper for the 5th International Congress on Large Dams in Paris in 1955 describing the case of Chew Stoke.

He did varying amounts of research on issues arising from each of the jobs in which he was involved and wrote nearly all of them up into papers for publication. It was this kind of work that gained him the London University DSc in 1949. He served on the Council of the Institution of Civil Engineers from 1949–54, representing the younger members, the AMICEs, and, between 1948 and 1957, he held a special lectureship at the Architectural Association.

Professor Brinch Hansen, Director of the Danish Geotechnical Institute, invited Skem to Copenhagen to give a lecture to the Danish Society of Civil Engineers. Skem and Nancy made plans to meet up with Bjerrum there and then travelled to Paris in 1952 for a European soil-mechanics conference with David Henkel, whose company they enjoyed. After the conference, they hired a small car and visited the great French cathedrals of Chartres, Sens, Beauvais, and Rheims. Henkel was detailed to drive, his first experience of driving on the right-hand side of the road, and he found the Étoile in Paris, with its *priorité à droite*, a nightmare. He has happier memories of a leisurely riverside picnic and all having a swim.

The Shellhaven cofferdam job of 1952 was notable for the first set of actual measurements in England of loads on struts in an excavation in soft clay, which was parallel to the work done earlier at Terzaghi's suggestion on the Chicago subway. At Tilbury the following year, Skem's studies gave the first information on the properties of peats and clays in the Thames marshes. This was presented together with David Henkel at the 1953 Third International Conference on Soil Mechanics at Zurich, in a paper entitled 'Postglacial Clays of the Thames Estuary at Tilbury and Shellhaven'. Much later, in 1995, in the context of the interest in the global rise in sea levels, Skem followed up the Shellhaven and

Tilbury work by studying the Quaternary geology of the West Tilbury marshes (Quaternary Research Association, Durham, 1995).

Zürich conference

The next International Soil Mechanics conference was in Zürich in 1953. By this time the International Society (ISSMFE) had become firmly established. Skem's paper was 'The Colloidal "Activity" of Clays'. He defines 'activity' as the approximately constant ratio of the plasticity index to the clay-fraction content in any particular clay stratum.

While in Switzerland for the Zürich conference, Skem took the opportunity of bus trips with Terzaghi, Casagrande, and others to visit Villa Olmo near Como. They also saw the beautiful reinforced concrete bridges built in the early 1900s by the Swiss architect-engineer Maillart. In his Tavenasa Bridge, built in 1905, the arch and the roadway, consisting of two slabs, one curved, the other straight, form one structural unity.

It was at the Zürich conference that Skem met the American engineer, Ralph Peck, for the first time. Peck was internationally known for his work with Terzaghi on the Chicago subway and also for their joint book, *Soil Mechanics in Engineering Practice* published in 1948, which had quickly become a 'main pillar of geotechnical education'. (Goodman). Since 1948, Skem had been bombarding Peck with requests for information about Chicago clay, for samples of thixotropic clay from New Haven, and for information about St Thuribe clay. He had also anxiously asked Peck about his views on clay sensitivity, perhaps double-checking in the light of Terzaghi's criticisms. Peck always responded diplomatically and at length, and with the eloquence and clarity of expression for which he is famous. The two men clearly had a similar approach to their subject. In reply to a query from Skem about a bearing-capacity equation, Peck wrote, 'I prefer to adopt the attitude that theories may come and go but the results of the field observations will retain permanent value, and should be tied to the theory as little as possible.' At Zürich, Peck told me, 'It was a late Sunday afternoon and I came to the Conference Hall to register. I noticed two men in animated conversation and recognized one of them as Skem (I had seen his photograph). I came up somewhat diffidently, waited for a pause, and introduced myself. His response was a somewhat tentative "Yes?" Then something clicked and he boomed out, "Oh, *Ralph!*" He introduced me to his companion, who was Laurits Bjerrum, then just finishing his doctorate at the Swiss Federal Institute and just appointed the first head of the new Norwegian Geotechnical Institute.'

Skem was very keen to get to know Peck and suggested that they all have dinner together. Bjerrum knew a congenial restaurant on the lake, and Skem ordered a bottle of wine – a good start to their friendship, as well as to the conference.

'That evening remains in my mind as perhaps the most enjoyable and fruitful I have ever spent,' says Skem. He goes on,

> Later, on the same trip, all three of us had another wonderful evening, with Hvorslev, as a result of which I determined to apply effective stress analysis to the London Clay slopes. In arriving at this decision, which at the time seemed almost far-fetched, I was encouraged, or even prompted, by Laurits. (*Géotechnique,* 1973)

Skem has a horror of making after-dinner speeches but, at Zürich, Terzaghi bullied him into it. Perhaps buoyed up by the euphoria of the whole occasion, he managed a speech that was a great success and he even dared to make good-humoured fun of Terzaghi. Peck obviously enjoyed it because he came up to him afterwards and said, 'A person who can give a speech like that can do anything!'

On his way back to the USA from Zürich, Peck stopped over in London, and spent more time with Skem. He remembers the first night of his visit, in the living room at The Little Boltons, with Skem, standing by the fire, Scotch and soda in hand, hotly debating with Silas Glossop, 'another tremendously enthusiastic man', the places he should see and the things he should do. Suggestions ranged over the entire technical history of England's cathedrals and industrial sights. They could even go as far as Land's End in Cornwall to see Trefethen's mine railways. Peck didn't get to bed until long after midnight, but no sooner had he dropped off than he was awoken at two in the morning by Skem knocking on his door. Skem apologized for the fact that he and Silas had proposed a much more strenuous agenda than could be carried out, and there was much they could see close to London. How about Stonehenge, and Salisbury Cathedral, 'one of Skempton's favourites of which his knowledge was intimate'? Philippa Wynne-Edwards (Skem's secretary), commandeered a small car belonging to Don Macdonald, a postgraduate student from Canada, and they all set off, plus picnic lunch, in sunny weather and had a wonderful time.

'Skempton is a physically big man; indeed everything about him is big – his smile, his laugh, his voice, his gestures. He even writes with a broad pen,' writes Peck. 'Whatever he did, whatever he discussed, was pervaded with an air of intense excitement. He never described large objects as merely large, they were *enormous*.' (Vignettes)

The tone of the correspondence between the Soil Mechanics section at Imperial College and Peck becomes much more familiar after Zürich. Philippa Wynne-Edwards, wrote him a hilarious account of life in the section in November 1953: 'We have a very mixed bag this year of postgrads – varying from a bright and breezy Canadian who talks endlessly but quite sensibly, thro' several Australians who talk endlessly and gratingly – Skem visibly blanches when he sees them – to two Indians, who I think must be mute. Oh, and a Turk who speaks no English at all as yet, but has a heart-stopping smile! Somehow we shall turn them out at the end of the year with some rudiments of Soil Mech.'

Sasumua Dam

Despite Skem's admiration for Terzaghi, he acknowledges that he was arrogant, perhaps particularly as he grew older. He saw himself as a giant in the profession.

In the mid-40s in Kenya, the Sasumua Dam, designed by Howard Humphreys & Sons for the City of Nairobi, was under construction and in dire trouble. The French contractor found that the clay being used to form the embankment not only contained an excessive amount of water, but also possessed peculiarly high values of the liquid limit despite low plasticity, and an exceedingly low density when compacted.

'For a particular intensity of compaction, each soil possesses an optimum water content at which compaction gives to the soil a maximum density. The experience of earthwork showed that an embankment constructed of soil wetter than its optimum water content might acquire both a low shear strength and abnormally high excess pore water pressures. This can be dealt with by flattening the slopes, slowing the rate of construction, or drying out the soil before it is compacted.' (Goodman)

It seemed that Humphreys would have to completely redesign the dam. He called upon the expertise of Guthlac Wilson and Skempton. Guthlac became a good friend of Skem's. He had studied soil mechanics at Harvard with Casagrande. He returned to England because of the war, and Skem had met him when he visited BRS to discuss his paper, 'The Settlement of London Due to Under-drainage of the London Clay.' He set up his own consulting firm, Scott and Wilson, in 1945 (and was later the structural engineer for the Royal Festival Hall). Skem admired him because he was by far the best consulting engineer who also had a detailed knowledge of soil mechanics. When he and his wife were tragically killed in a plane crash in Africa in 1953, Skem wrote his obituary in *Géotechnique*. After detailing his engineering achievements, he wrote, 'He was a delightful companion with an exceptionally wide range of interests and accomplishments.'

Skem undertook mineralogical tests on samples from Sasumua and obtained a mass of data. He found that the material was formed primarily of an unusual mineral, the clay halloysite, whose particles are not platelets like those of most clays, but hollow tubes. In this soil, water occupies not only the usual sites between the particles of clay, but also within the tubular particles. Since this extra water lies inside the tubes, it cannot influence the forces between the particles and is, therefore, inactive with respect to the behaviour of the clay. In the 1998 interview, Skem told Dick Chandler that the properties of the clay 'were wildly disparate with any code of practice you could think of, or for that matter anything in any textbook. I think it was the first time that such a material had come into any major engineering job.' Wilson and Skempton also suggested a redesign. The City of Nairobi declared the contractor to be in default and engaged Terzaghi to advise. Terzaghi spent a day or two at Soil Mechanics Ltd in Chelsea. He confirmed Skempton's findings and arbitration hearings were held. (By the time these took place, in 1953, the dam had been built by the Nairobi City engineer, and the reservoir filled.) These hearings are notorious for the manner in which Terzaghi behaved under cross-examination. Skem, who was present giving evidence about the properties of soils, describes the painful scene:

> The lawyer for the opposite side was Salmon, who became Lord Justice Salmon in due course. He was one of those barristers with an absolutely razor-sharp mind and with no respect for anybody. He obviously took a dislike to Terzaghi, who I must say gave a rather arrogant impression. He was doing the Great Man act, when it wasn't called for. Salmon demolished him.

1953 was Coronation Year. Skem marked the occasion, as did so many other people in the UK, by purchasing a black and white television so that Beatrice could watch it. After this, sunny weekend afternoons at Watford were passed with the curtains tight closed so that the screen became visible, as Skem and Beatrice sat glued to the test match.

Back at the Rotterdam conference in 1948, in one of his many papers, Skem had attempted to rationalize the subject of field findings on shear strength, in the so-called lambda theory. Ralph Peck writes that this was 'enlightening but not of practical use because the necessary physical properties could not easily be evaluated. However, in 1954 he published the paper, 'The Pore-pressure Co-efficients A and B' in *Géotechnique*, that immediately clarified and simplified the whole subject of the shear strength of saturated soils. Although less than five pages, it caused great excitement in the soil-mechanics community and redirected research and understanding of shear strength for a generation. Imperial College quickly became the center of shear-strength research and teaching.' (Personal communication)

Travels in America

That same year, 1954, was a time of extensive travel for Skem and Nancy. Ralph Peck invited Skem to America to give a series of lectures in different universities across the States. He did an enormous amount of work drumming up support for the tour and finance for speaking engagements, so that, in those days of austerity in the UK Skem could afford to bring Nancy with him. His efforts bore fruit and, in April, the couple set sail from Southampton for the four-day voyage to New York on the Cunard Line *Queen Mary*, even then the grand old lady of cruise liners. Skem enjoyed her comfort, the spacious cabins, and the lack of roll. Huge waves seemed to make little impact on the stability of the ship. His album notes: 'Arrived New York April 27th, speed 28 knots'. After a few days' sightseeing in New York they moved on to Boston, to stay with Arthur and Helen Shaw, who were friends of Beatrice from her Clinton, Mass. days. (Skem remembers them coming to England when he was a schoolboy and him and his mother going to stay with the Shaws in a hotel in Piccadilly. When Skem first went to BRS, Mr Shaw who was also, coincidentally, an engineer, had sent him some papers from the *Journal of the Boston Scoiety of Civil Engineers*, a journal which published some of the most important early papers on civil engineering from Harvard and MIT.) They travelled together to Beatrice's favourite, Cape Cod, where they stayed in a motel in Falmouth and were particularly struck by the beauty of Sandwich. They took tea in the very stylish Boston Ladies' Club, where the women talked in cut-glass South Kensington accents, with only slight American inflections. Nancy and Skem commented on this, and the Shaws confirmed that this is how upper-class New Englanders like to speak. They then went to stay for ten days at Karl and Ruth Terzaghi's spacious house in Winchester, a satellite town of Boston. Skem remembers that, during their visit, Karl used every now and then to retreat to his upstairs study to work and chain-smoke and later come down, quite refreshed. Karl was teaching at Harvard, and Ruth drove him the ten miles to his office at the university every day.

(At the beginning of his career at Harvard, Arthur Casagrande had been Terzaghi's assistant. He then ran the course very capably for several years while Terz was in Europe. When Terzaghi came back to Harvard, he was given an office right next to Casagrande. This must, Skem thinks, have been 'rather a trial' for the naturally rather hesitant and indecisive Casagrande. 'The rest of us only had to put up with Terz for a week or two a year, but Casagrande had him looming over him every single day.')

While Skem gave some lectures, Ruth and Nancy explored Boston. It was a thoroughly enjoyable visit, due in large part to Ruth's hospitality. From there, they went by train to Urbana, Illinois, where they stayed with Ralph Peck, his wife Marjorie, and their young children, Jimmy and Nancy, who gave up her bedroom for the Skemptons' use. The *News Gazette* for May 9th announced that Skem would give a series of eight lectures, four on soil-foundations problems and three on the early history of modern civil engineering. 'In addition, Dr Skempton will address the central Illinois section of the American Society of Civil Engineers on the structural development of the medieval [sic] cathedral.' On the 13th, the Pecks drove Skem to Purdue where he lectured to Jerry Leonards' students on shear strength and, on the 18th, they all visited Abraham Lincoln's birthplace and the log cabin, where he had been postmaster from 1833–36 and other scenes around central Illinois.

On their way home, they stopped at an old farmhouse converted to a restaurant, noted widely for its country-fried chicken dinners. As was the custom, the chicken was presented in pieces in one large basket in the centre of the table. Each place setting had a plate and serviette, but no cutlery. Peck remembers, 'Marjorie and I realized that Nancy and Skem were trying to figure out how to eat the chicken; they saw we used our fingers and, after considerable hesitation, decided to go along with these uncivilized ways of England's one-time colonials.'

A more important aspect of the trip for Skem can be deduced, however, by photos he took of a landslide in the 'Wisconsin Loess' and 'Kansan and Nabraskan till' in a railway cutting somewhere in Iowa. Skem was amazed at the huge agricultural landscapes of America, every inch cultivated, the silos and enormous barns, so different from the English shires. They then travelled yet further west to Denver and, after lecturing at the University of Colorado and parting company with Peck, motored high into the Rockies through the Berthoud Pass and up to the Granby Dam with Wes Holtz and Fred Walker, the designer of the 250-ft high dam. They took the 'Streamliner' train all the way from Denver back to Chicago, where they re-joined Ralph Peck, and Skem admired the iron-frame buildings such as the Marshall Field retail store. Peck describes their explorations in the city: 'He had long been intrigued with the transition from masonry structures to the modern framed building. Part of the development had occurred in England, but some of the more significant advances were made in Chicago in the late 1800s. Several famous old buildings were still standing, others had recently been destroyed. Skempton was keen to see the survivors before it was too late… In the company of Sidney Berman, who was then in charge of soil mechanics for the Chicago subway and who had a genius for enlisting the cooperation, perhaps unwitting, of custodians and security guards, we dashed in and out of the basements, stairwells, and elevator shafts of many an old structure, some almost ruins.'

The newsletter of the American Society of Civil Engineers Soil Mechanics and Foundations Division for June 1954 sums up the demanding schedule of lectures that Skem set himself, and it also seems that Nancy made quite an impact as well. 'In the course of his travels, Dr. Skempton is delivering lectures on a number of subjects of current interest, including shear strength, consolidation, bearing capacity, and sensitivity of clays, stability of slopes and earth dams, and the history of civil engineering. These talks, to date, have included one or more at Harvard University, Massachusetts Institute of Technology, the Boston Society of Civil Engineers, the Bureau of Reclamation, Purdue University, the University of Illinois, and the Geotechnical Center of Northwestern University. All the lectures have been well attended and greatly appreciated by the audiences. Dr. Skempton is an excellent lecturer; much of his material has been new, and all of it has been interesting and stimulating. Everyone who had the pleasure of meeting Dr. Skempton and his charming wife looks forward to meeting them again.'

Next stop was Toronto, and then Niagara Falls, where Skem photographed the cofferdam for intake works at the Sir Adam Beck Niagara Generating Station No. 2, and the 18th-century Fort St George, which he visited with his research student, Don Macdonald. Don was one of a number of Canadians who came to IC on Athlone Fellowships. These were funded by the Earl of Athlone, with the idea that exceptional students would undertake postgraduate studies in England, and take their skills back to Canada. Norbert (Nordie) Morgenstern was a later Athlone scholar arriving in 1958. Don became particularly close to Skem and our family, and Katherine remembers flirting with him on his visits to The Boltons. Before his lecture in Ottawa on May 31st, they drove along the St Lawrence and Rideau Canal. Then another lecture at Montreal, before they sailed away from the New World on the Cunard White Star liner *Ascania*. They were much delayed by icebergs and bad weather in the Atlantic. The foghorn blared night and day and Skem got thoroughly bored, despite an invitation to cocktails with the *Ascania*'s Captain, E. Divers OBE. They were finally piloted into Liverpool docks on June 11th 1954, just in time for the summer holiday in Yorkshire.

Scarborough

Skem and his family spent three weeks or a month every year on holiday with Nancy's parents in Scarborough and this was a very important part of Skem's year. He truly relaxed in Yorkshire. Reginald and Rosa led a life of strict Edwardian daily routine. Breakfast was at eight, then Rosa and the maid, Eva, spent the morning attending to housework and laundry, the process that Skem remembers bore resemblances to Mrs Stubley's labour at Northampton. Rosa would roll up her sleeves, don Wellington boots and a tweed apron, and go into the quarry-tiled scullery, where there was a china 'butler's sink' with a grooved wooden draining board. Like Mrs Stubley in Northampton, she used a boiler, a three-legged wooden 'dolly' and a washboard. Jim, the gardener/chauffeur, got out the four-by-four posts, slotted them into their post holes, and slung the washing line all round the garden. From midday onwards, the garden became impenetrable unless you could run the gauntlet of huge damp linen sheets or Reg's extraordinary underwear flapping in your face. Meanwhile, he did the shopping in the black Rover or picked flowers, which he arranged at the butler's sink and set around the house in beautiful cut-glass vases, or collected fresh beans or raspberries from the garden for lunch, which took place on the dot of one o'clock. Woe betide anyone who was late.

One morning Skem was out measuring some geological formation in the neighbourhood and, realizing he would be late, drove the car 'like a madman' through Scarborough. He arrived at the lunch table panting but Rosa barely spoke to him for two or three days, in such bad odour was he. At the meal, Rosa would summon Eva between courses with a little bell, which she kept on the white damask tablecloth. Eva served the vegetables in elegant dishes and Reg rested the carving knife on cut-glass supports, while the sun shone down on the rose garden outside the dining-room window. Rosa then retired to bed until about four o'clock, when she emerged for afternoon tea. With our visits, a parallel family routine was quickly established, as Rosa was one of the few people to whose demands Skem submitted himself. Every morning he would disappear up to the large bare workroom with his pipe and whatever project he was working on. His journal for 1947 notes, for example, 'July 24th – August 18th Holiday in Scarborough. Wrote introduction to the Cornell book on Collin'. He would take a break for morning coffee and biscuits. Meanwhile, Katherine and I would play on the swing hanging from the cherry tree or make miniature gardens in trays, using materials found all over the garden. Silver paper from sweet wrapping made the ponds.

After washing up the lunch dishes (Eva had the afternoon off), Nancy would pack a picnic tea and a family expedition would be undertaken to one of a range of beauty spots on the North Yorkshire Moors – Hutton-le-Hole, or Silpho Moor. Or we would go to a beach, at Cornelian Bay or Robin Hood's Bay, where we would join up with Nancy's sister Betty and her two children, Brenda, who is my age, and Richard who is Katherine's.

Skem loved going to the local village agricultural shows at Burniston or Thornton-le-Dale and enjoyed the rural pursuits of sheepdog trials, the latest advances in tractor design, seeing who has the biggest marrow or French beans in the marquee, and watching farmers' daughters showjumping on their ponies. Sometimes we would venture away from the Wood in-laws and spend a few days in more remote Pennine villages, for example, Goathland in 1951 or Arncliffe in 1953.

NGI – Laurits Bjerrum

Immediately after the 1954 summer holiday, Skem travelled with Nancy to Oslo to visit the Norwegian Geotechnical Institute, where Bjerrum had just been appointed as the first (and, according to Skem, 'inspired') director. It became one of the leading research institutes in the world, and the Norwegian equivalent of BRS. Skem found it a tremendous place, buzzing with work and new ideas. A geological phenomenon that is more common in Norway than elsewhere is the 'flow slide' in postglacial clay (that is, clay that was deposited under the sea in the late glacial period). As the Ice Age ended, the land had risen due to the melting of the ice. Rainwater then leached out the salt, leaving the clay in a supersensitive condition. Small slides are initiated, which result in the clay liquefying. Bjerrum and his colleagues had discovered the reason for these slides. It was very exciting for Skem to discover these parallels to his own work on postglacial clays in England and to find such a kindred spirit in Laurits. The Skemptons stayed with Bjerrum and his wife Gudrun and went on sightseeing trips together. Bjerrum and Skem were as close as Skem got to anyone. Skem responded to the vivacity and energy of Bjerrum. He very much admired his intelligence and considered him an exceptional scientist. Bjerrum grasped and analysed technical problems very quickly and asked the right questions. He could be very firm with his staff and would take the lead in any situation. He was perhaps more status-conscious and ambitious than Skem, qualities which may have contributed to the very success of the Institute. NGI research and methods for solution of problems in connection with constructions in soft clay are now incorporated in textbooks in use across the world, and have given Dr Bjerrum and NGI a respected international reputation.

Skem and Nancy went on to Sweden, to the European Conference on Stability of Earth Slopes in September 1954, where, at the final dinner, it was Nancy's turn to give a speech, probably thanking the conference organizers on behalf of the ladies. They visited the Swedish Geotechnical Research Institute at Uppsala. The highlight of the visit for Skem was seeing the Handel opera *Orlando* performed in the beautiful baroque theatre at Drottningholm Palace with 18th-century stage sets and *deus ex machina* effects. This was the first Handel opera he had seen. They were rarely, if ever, performed in London at that time but the Drottningholm experience led Skem to Sadler's Wells, where the Handel Opera Society was making tentative efforts at revivals of these wonderful works.

The musical engineer

At this rich and full period in his life, Skem's musical interests burgeoned. Through Julian Trevelyan, Skem met Peggy Spring-Rice and her circle. Peggy played the flute enthusiastically but 'very breathily', according to Skem. In the late 40s, this group, including Rosamund Jenkinson, a teacher of English at St Paul's Girls' School, and a keen oboeist, met as a chamber group in the house of some friends of Julian's in Brook Green, Hammersmith. There was a large room graced with *two* grand pianos. Skem was friendly with another good amateur flautist, Julian Towers, who lived not far from Ursula and Julian in Chiswick. His sister sang and a friend of the sister played the piano. During the time of the chamber group, Skem had his second set of flute lessons with Fritz Spiegel. At home as a child, I remember going to sleep to the sound of my parents playing flute and piano sonatas. When Katherine was old enough to learn the cello, she was pressed into service and, to her terror, had to take part in musical evenings of baroque trio sonatas. Boismortier, she remembers. Skem could play them well – he practised – but, if the piece had more than two sharps, poor Katherine was stumbling.

Peggy Spring-Rice, who was a friend of the composer Ralph Vaughan Williams, got Skem an invitation for the first time in 1947, to join a group of friends and admirers of Vaughan Williams at his Leith Hill Music Festival, which had continued throughout the war. Vaughan Williams (RVW) was a passionate advocate of amateur music-making. Skem played second flute in Bach's *St Matthew Passion* at Redhill and Bury St Edmunds and then in the *B Minor Mass* at Dorking Halls in April. In Skem's journal is pasted a thank-you

Chamber music at Silwood with Skem on flute.

letter from RVW, who comments 'As many flutes as possible should play in the *ritornelli* of the *Domine Deus*... you five sounded fine and all in tune!' The *St Matthew* became an annual event. These were ground-breaking but deeply unauthentic Bach performances. Vaughan Williams had his own version of the monumental works, sung in English, with huge musical forces, drastic cuts, and a romantic interpretation and he conducted with idiosyncratic tempi. The other flautists in these performances were top professionals brought in by RVW, such as Geoffrey Gilbert, Gareth Morris, and Gerald Jackson. Skem always laughs when he remembers a rehearsal on a particularly cold spring day in the halls. Jackson wore navy-blue fingerless woollen mittens and, when a particularly fast and technically difficult flute section, the *Moon and Stars,* came along, he calmly removed his gloves, rattled through the passage, and then put them on again. Ever after, Peggy and Skem called that the 'gloves-off' passage. Peggy Spring-Rice, in turn, introduced Skem to Dr Fritz Arnholtz, a Jewish refugee from Nazi Austria, who impressed Skem with his large collection of books. He was, for many years, the Skempton family's GP in London. He was tremendously grateful to Skem for getting him a reader's ticket to the British Museum, which seemed to him to symbolize acceptance into London intellectual life.

Another outlet for Skem's chamber music talents was the Imperial College lunchtime concerts. The musical life of the college was initiated by a small group of keen staff and students, aided and abetted in every way by the then Rector, Sir Roderic Hill. The City and Guilds Council Room was made available and was ideal, providing seating for an audience of 100 or more in gracious surroundings. As Harold Allen remembers, 'Here, in the summer of 1950, David Tombs launched the series of weekly lunchtime chamber concerts that has continued to this day. Initially, most of the performers were members of the College (students and staff): they included Harold Allan (violin), Eric Brown (piano), Alec Skempton (flute), Bryan Thwaites (piano and bass voice), and David Tombs (tenor). Musical friends and students from the Royal College of Music soon joined in.' Silver collections were taken at the end to defray the costs. Skem's album contains several programmes from the concerts in 1950 and '51, of Bach cantatas and Handel trio sonatas, and, in addition to the above performers, Nicola Darwin is often mentioned as soprano. Skem himself wrote the programme notes for the concert on February 15th 1951, of music by Haydn, Handel, Mozart, and a certain Schmelzer, who, Skem informs us, was *Kapellmeister* at the Imperial Court at Vienna. 'This unpretentious sonata was a piece for a group of students to play on seven recorders or any other instruments that happened to be available.' The only photo I have of Skem playing the flute is with the IC group in June 1952 at Silwood Park, Imperial College's field station.

Although Skem can be prone to outbursts of anger himself (he has been known to exclaim, 'That is un-called for', or, worse, 'The man's a s..t'), he has a deep distrust of any outward show of emotion or feelings of any kind in other people. Bob Gibson tells a story about one of the Imperial College lunchtime concerts, which illustrates Skem's difficulty in dealing with and avoidance of emotionally delicate situations. On this occasion, the ensemble was a pianist, a cellist, and a single French horn. The horn player was struggling with his instrument, kept on blowing raspberries, and eventually went berserk. He threw his horn into the audience, where it landed in the lap of an astonished Philippa Wynne-Edwards, stormed over, and banged on the piano, screaming at the top of his voice. It was disturbing to witness someone losing control so completely. The musicians gradually pulled themselves together and hesitantly continued playing. Skem was the most senior staff member present, but it was Eric Brown of Structures who was left to resolve the situation.

Skem made his own small contribution to musicological research by publishing a paper in the journal *Music and Letters* (in 1962) on the instrumental sonatas of the Loeillet family of French baroque flautist/composers, whose life and work he had researched as thoroughly as any landslip or raised beach. (He also wrote the article on Loeillet in the *New Grove Dictionary of Music*.)

Meanwhile, Katherine had followed me to St David's and I had passed the frightening entrance exams to St Paul's Girls' School at Brook Green, and was having piano lessons there. My teacher had the engaging name of Ruby White, who always had an extraordinary number of shoulder straps showing through her pale nylon blouse. I quickly learned that, if I did my piano practice after supper, I was not asked to help with the washing-up. Skem made one of his very rare visits to the school at the end of the autumn term 1954 to hear me, a nervous twelve-year-old, play Bach's *Gavotte* at an 'informal concert' in the music room.

Postal history

Another interest of Skem's, much more of a 'hobby' than the rather serious concern with music, was postal history. In the early '50s, he had built up an extensive stamp collection but he sold this for a few hundred pounds in order to raise money to buy whole envelopes, with their hand stamps and postmarks and evocations of an earlier age, distant people and places, and the days before the official post-office system. Old solicitors' offices were the main source. Skem would regularly get a shoebox full of old letters from his dealer, Willcocks of Blackheath. They were mainly from the 18th century, though he has some 17th century ones. Each envelope cost sixpence or a shilling. He would select about twelve from each box and return the rest. The presiding genius of the postal history world was Robson Lowe of 50 Pall Mall, later taken over by Sotheby's. Skem went to a few auctions between 1952 and 1960. He also went to meetings,at a hotel in Bloomsbury,of the Postal History Society, annual subscription three guineas a year.

On January 13th 1955, a small notice appeared in *The Times* announcing that Dr A.W. Skempton had been appointed to the University Chair of Soil Mechanics. This was a personal chair created for him by Pippard, the first in soil mechanics in the country. The next chapter in the story of the development of the subject and Skem's professorial career at Imperial College, was about to begin.

Chapter 5

Move to The Boltons– Head of Department 1955–60

There were nine months between Skem's appointment as Professor of Soil Mechanics and his inaugural lecture, a nine months packed with activity. During the 1955 Easter holidays, he and Nancy had a short break staying at The Lamb at Lavenham in Suffolk and visiting the beautiful East Anglian 'wool' churches, including Long Melford and Thaxted.

Paris Congress on Large Dams

In June 1955 came the Fifth International Congress on Large Dams, in Paris, where Skem and Alan Bishop presented their work on the Chew Stoke Dam, where they had studied pore-pressure dissipation in soft clay foundations and installed sand drains for accelerating consolidation in the soft foundation layer (see pages 87–8).

It was probably at this Paris conference that Skem met the cultivated and eminent French engineer J.L Kerisel, of the École des Ponts et Chaussées. Skem and Nancy admired and quickly became close friends with the Kerisels. Madame Kerisel was the daughter of the distinguished Inspecteur Général des Ponts et Chaussées and member of the Académie Français, A. Caquot. Skem gained some credibility with Caquot and Kerisel because it had been he, as far back as 1946, who had rediscovered the 19th-century French engineer, Alexandre Collin. He had published 'Alexandre Collin (1808–1890), Pioneer in Soil Mechanics' in the *Transactions of the Newcomen Society* of that year, a paper which Kerisel translated into French in 1956. As Kerisel wrote (personal communication), 'Collin... had totally fallen into oblivion when rediscovered by A.W. Skempton... Collin was the first to measure the shear strength of clay samples, using this strength in an analysis of stability. Above all, he humbly recognized that satisfactory solutions to the various problems of soil mechanics must "one day be the reward of those who, without separating mechanics from natural philosophy, are able to correlate the principles of the former with those facts which it is the purpose of the latter to discover and coordinate."' (Personal communication)

It is Skem's view that, in France, engineering, having an altogether higher status than in the UK, attracts a higher calibre of better-educated people than is the case in the UK where the brightest young people tend not to choose engineering as a career.

After the Paris conference, Skem and Nancy made a trip to explore the early French canals, visiting the Orléans Canal and then the Canal de Briare, south of Paris, the first great canal in Europe, built in the early 17th century. Kerisel kindly lent the Skemptons his own car for the canal expedition.

There was also a great deal of touring around during that year's summer holiday in Yorkshire. They went as far north as Barnard Castle and spent a week in Arncliffe. They must have got their own car by then, perhaps the Morris Oxford that Skem bought from Don Macdonald.

Skem's inaugural professorial lecture addressed the appropriate subject, 'Soil mechanics and its place in the university.' In preparation for this, he wrote asking Peck about the number of full professorships there were in the United States at that time and when they were created. His lecture was not the big event that such speeches have nowadays become. Skem thinks they are 'tricky' things. They are attended by the rector, your own colleagues, various diligent members of other departments, and wives and friends. Striking the right note in an inaugural lecture is not easy, as it must steer a steady course between technicality on the one hand and banality on the other.

He ended the hour-long lecture quoting two lines from William Blake,

To see a World in a grain of sand…

And Eternity in an hour.

'The first, I trust, may be considered apt: the second, I hope is not true, at least so far as my audience is concerned.'

He returned to this quotation years later when awarded the Gold Medal of the Institution of Structural engineers, with an addition: '"To see the World in a grain of sand" or in my case a Particle of clay…'

Carlotta

Philippa Wynne-Edwards departed for South Africa with her new husband, and the red-headed talented English graduate, Carlotta Hacker, then in her early 20s, joined the department as Skem's right-hand woman. The Eastwick-Fields had told him about her. She gave me a good example of how Skem looked after the people who worked closely with him, and how he created such a cohesive group of colleagues in the Soil Mechanics section. 'The first time he gave me a paper to type, he asked me if I could do it in four days. When I turned it in after three days, he was absolutely delighted, thanked me effusively, so of course I glowed. And next time I had a paper to type, I tried to do it even better. It was the same with any research I did for him. Lots of warm thanks and praise. Yet his praise wasn't calculating in any way. He thanked people and praised their work because it was his nature to do so.' She not only respected him, she was fond of him. 'Being in touch with a mind like that every day must have shaped my own development far more than I realized at the time,' and despite the age difference, they discussed things intensely. She not only typed, seemingly for the whole section, but Skem, seeing her potential, gave her 'the most superb training' as a researcher. 'The first time he asked me to look up something in the British Museum I came back to him as pleased as Punch because I had found what he wanted, and had copied it all out. It included a reference to an earlier work, one written in the 1740s, I think, and Skem asked what it had said. Oh. I hadn't thought of following the trail back to its origins. Skem then suggested – surprisingly gently, considering how dumb I had been – that I go back next day and check out the 1740s book. Well, of course, by the time I'd traced the story back to its source through several books, it was significantly different from the book I'd originally looked at.' There was definitely a sexual spark between Skem and Carlo, which many colleagues have remarked upon.

Staff group

The small Soil Mechanics staff group still consisted of Skem, Bishop, Gibson, and Henkel. Nick Ambraseys, a Greek engineer, who arrived as a postgraduate student in 1955, characterizes the three in this way: 'Bishop was the lab man, the brilliant, difficult one; Henkel was the field man, who would call a spade a spade; Gibson was the theoretician, the well-spoken, quiet, perceptive one, who would always do a bit more than he was asked to do. They complemented each other.' Skem gave the lead to the three. The department was blessed with a pleasant coffee room with newspapers and journals, *The Times*, and *Nature*. If a joint paper was in the writing, people would meet in the coffee room at lunchtime, talking 'shop'. It was an opportunity to meet people from other parts of the department. If it was fine weather, Skem would suggest to Bob Gibson, who was not a great one for exercise, 'Let's go and have a walk round the park.' 'Sometimes we would find ourselves going twice round Kensington Gardens, if the talk went on, before finding ourselves back in our rooms,' says Ambraseys. The walks were an excuse for Skem to think aloud. He did not always listen to colleagues' replies and this 'switching-off' of Skem's can be unnerving. Ambraseys feels it is inherent in the way Skem works. When he has collected enough information on a topic into his 'databank', he is simply not interested in discussions with people who have little to add.

David Henkel

Henkel had uprooted himself from his country of birth, South Africa, then a dangerously racist society and, as a 'colonial', was somewhat of an outsider in the rather closed world of the Soil Mechanics section at Imperial College. Bob

Gibson characterized Henkel as a very good 'all-rounder' but without either the excellent experimental skills of Bishop or the mathematical ability of Gibson. Skem describes a certain amount of rivalry between him and Bishop, who got his readership, and subsequently his professorship, ahead of Henkel. Bishop and Henkel together wrote the classic text, *The Measurement of Soil Properties in the Triaxial Test*, published in 1957. This is, in fact, one of the few books to come out of IC during Skem's reign. Skem set an excellent example of writing up case records but never wrote a soil-mechanics book. When I asked Skem why he had never written, for example, a textbook of soil mechanics, he replied that he was always too much interested in the ongoing research problems arising out of each job that he was undertaking, in finding solutions to real problems, to devote the necessary years to writing a book, which, in any case, he feels would soon be out of date, as the subject was advancing so quickly at that time.

Henkel would arrive each day at college on his bike, which was quite a spectacle, as he is a heavily built man. He and Skem spent many enjoyable days together studying stability of slopes, looking at slips in London clay, including a steep clay slope on Sydenham Hill, SE26, where Ove Arup had built some flats for the LCC, where gravel drains were needed. They also investigated clay in a railway tunnel in the Avon Gorge in Bristol. Henkel remembers travelling down with Skem from Paddington to Bristol for day trips on the 'Bristolian' steam train.

Jackfield

One of Skem and Henkel's most significant studies in the early 1950s had been at a landslide at Jackfield, just downstream from Ironbridge on the banks of the River Severn in Shropshire. The resulting paper had made an immediate impact. Skem and Henkel applied an effective stress analysis based on measured pore pressures and drained shear tests, using geometry defined by the observed slip surface. 'They found the back-analysed factor of safety to be too high by nearly 50 per cent, reasonable agreement being obtained only when the cohesion intercept was neglected... This seemed to be consistent with the observed much higher water content on the slip surface, as compared with water contents in the clay elsewhere.' (Selected papers)

Henkel described to me Skem's way of working: 'It was his careful observations and detailed measurements, often with the simple Abney level, which were typical of his methods of working and from which I learned a great deal. The details, however, were never allowed to dim his concern for the overall picture and how the particular fitted into the geological environment and history. The study of the stability of the Monar Dam in Scotland was also an exciting experience (for us) in complex geology.' He found his time in the Soil Mechanics section intellectually very exciting. He sums up the work of the section as providing the experimental back-up for and exploration of the consequences of Terzaghi's principle of effective stress. 'In this work, Skem provided inspiring leadership,' he writes. Henkel was an inspiring teacher, and it was he who introduced the field trips for postgraduate students but colleagues report that he could be a little tetchy – you never quite knew what to expect. Staff felt, after a clash with him, as if they had been 'Henkeled'. He moved on to work in prestigious posts at the Indian Institute of Technology, where he battled with bureaucracy and indeed corruption, and at Cornell University, New York, and later came back to England to a senior position with Ove Arup. He was the Rankine lecturer in 1982.

In the volume of *Selected Papers on Soil Mechanics*, assembled on the occasion of Skem's 70th birthday, Bob Gibson gives a vivid picture of why working with Skem was so rewarding for staff and students alike. He has 'an outstanding ability to reduce a problem to its essentials. A stimulating discussion invariably follows in which everyone joins: this seeks to identify the important questions to be answered and considers how best to go about this task. Understanding the need for opinions to be expressed freely, he guides discussion with a light touch, regarding himself merely as *primus inter pares*. Irrelevant remarks are allowed to pass; foolish suggestions call forth only mild disagreement; clever ideas are welcomed, but not unduly praised. This style reflects Skempton's innate consideration for others, his powers of judgement, and his determination, rightly, to reserve for himself the final word. Conclusions are summarized succinctly and what needs now to be done is stated unequivocally.

'When he anticipates unusual difficulties, this stage may be followed by consultation with those whose knowledge in special areas he respects. Never content merely to accept what he is told, he adopts the mantle of the student and questions them closely to reveal the path along which conclusions have been reached and, furthermore, to master for himself the details of the reasoning. He will not hesitate to acknowledge his inability initially to follow an argument but will persist until its essence has been grasped and he has formed his own opinion. This unassuming and scholarly approach, entirely free from pomposity, makes a profound impression on his students and they, of course, warm and respond to it.'

Alan Bishop

The department at IC became famous for providing the experimental back-up and exploration of all the consequences of the principle of effective stress (see Chap. 3 p. 36) and having it accepted by the engineering profession at large. Skem provided the leadership but he pays generous tribute to Alan Bishop's contribution. Bishop, says Skem, 'never stopped thinking about effective stress. It wasn't exactly an obsession, but it dominated his life. He had it straight in the front of his mind, as a conscious objective. His very logical mind dictated that unless you were using the principle throughout, you were not doing the job properly. He showed how the principle could be brought into all sorts of practical and laboratory problems. In seven years he turned the subject round. He ended up with methods of designing earth dams using the principle of effective stress that are now taught throughout the world... I adored his ruthless intellectual rigour.' Skem continues, 'You couldn't slip anything past him. He was aware of what you were going to say almost before you said it. This did not endear him to people. He wasn't terribly liked.' Bishop led the tribunal team of investigators into the cause of the disastrous Aberfan colliery tip slide in 1966.

Although Bishop lived during the week in the somewhat austere Quaker Penn Club in Bloomsbury, he spent every weekend at his parents' house in Whitstable, Kent, where he had a sailing boat. It was a converted whaler, which had, in the past, been carried by Admiralty vessels as a lifeboat. It was about 35 feet long and clinker-built. Bishop had had a cabin built for it and mast and rig fitted. He took colleagues out in this and these trips were differently experienced by participants. David Henkel remembers a nightmare experience in the Thames estuary. They were sailing round Canvey Island when a huge oil tanker bore down upon them. Bishop refused absolutely to go about, maintaining that the rule was that the tanker should give way to sail. On another occasion, Bob Gibson and his wife, Elizabeth, felt so badly treated by Bishop during a voyage on the Essex waterways that they left prematurely. Bishop could not cope with their dog and also demanded that they go and get provisions, for which he never paid them.

Bishop was not a morning person. He never started work until 11 a.m. and worked very late at night. He put in a full day but a few hours out from the rest of the staff. Penman, who worked with Bishop on several dams, remembers that the normal time to get to a site investigation would be 8 a.m., which meant being picked up at 7.15 a.m. Not Bishop. He would say, 'Oh, that's too early, pick me up at quarter to eleven!'

Gibson felt Bishop never appreciated other people's circumstances. He would say, 'Oh, Bob, could you just help me tonight. I must complete this experiment this evening and all my research students have gone home.' When Gibson, who had young children, protested that Elizabeth would have prepared his evening meal and that he had a long journey, Bishop would say, cavalierly, 'Oh, just go and phone her and let her know you'll be late.'

'He was not one of us,' says Gibson. He didn't drink or smoke (or even less have sexual adventures). He had, Gibson said, 'a wet handshake'. Bishop paid a price for all this intellectual rigour and human rigidity. After the death of his mother, he suffered a serious breakdown with acute depression, and spent many years in psychiatric hospital, for some of that time confined in a straightjacket. He eventually recovered enough to return to college but he was a shadow of his former self, both physically and intellectually, and caused consternation among colleagues by taking an unconscionable time to produce one paper. He took early retirement, and, to the astonishment of his friends, who had always seen him as a confirmed bachelor, he married late in life and found contentment

Bishop – 'You couldn't slip anything past him.'

in his garden. Across the world there is a small, devoted, and extremely well-educated group of his ex-research students. Bob Gibson was touched that, when Elizabeth died, Bishop phoned and suggested he come up to their cottage in Scotland, in case he might be lonely. When he died in 1988, his widow arranged a quiet funeral and he was never properly honoured by the soil-mechanics fraternity. Skem added a footnote to the obituary in *Géotechnique*: 'In no respect was [Bishop's] intellectual power seen more clearly than in his continued study of Terzaghi's principle of effective stress and its application in all branches of geotechnical engineering; his work in this field brought about ... revolution in soil mechanics.' (*Géotechnique,* Vol. 38, p.653–55). Skem is upset that he never became a Fellow of the Royal Society, which he fully deserved, acknowledging that his rather obsessive personality might have acted against him in the fiercely competitive selection procedures.

Don Macdonald

Much of 1955 was taken up, for Skem, with work with his exceptionally able postgraduate student, Don Macdonald on their paper for the Civils, 'Settlement Analyses of Six Structures in Chicago and London'. Don's PhD thesis formed a basis for this. (Nancy bound a copy of the thesis for Ralph Peck.) Skem's involvement in this was connected to his fascination with the Chicago foundations, which he had viewed with Peck during his American tour the previous year. Skem and Don were again bombarding Peck with requests for further details of the foundations of buildings they had seen and drafts of the paper, to the extent that Skem included Peck in the authorship. Peck commented, amused, 'I may say that I can hardly imagine an easier way to write a paper than to have two other fellows do the work.' (letter 10.2.55)

In September, there was a conference at the Institution of Civil Engineers on 'Correlation between Calculated and Observed Stresses and Displacements in Structures'. Skem and Don gave a paper in the fourth session on earth pressures and movement.

In October, Peck wrote consoling Skem over another intemperate Terzaghi attack, this time over another of his papers with Macdonald, 'The Allowable Settlements of Buildings'. Terzaghi had written, 'My principal concerns center about [sic] the influence of the soils profile on the ratio between angular distortion and maximum settlement... the overwhelming majority of your readers will accept your suggested design limits with a sigh of relief... These probable consequences of the publication of your paper disturbed me so profoundly for weeks in a stretch. Presently I am trying to get over it.' (letter 7.10.55)

Peck had learned to survive such attacks, and consoled Skem with another letter: 'One of the rewards of association with Terzaghi is getting letters like the one he wrote you. I have been getting documents of that sort off and on for about ten years and, after the first shock, I always find them exhilarating and beneficial as well as chastening... Terzaghi would not take the time to write such a letter if he did not have a deep personal regard for the recipient.' (letter 16.10.55) Especially late in life, Terzaghi was sometimes prone to overstatement and intemperate outbursts. Skem now regards this kind of comment from Terzaghi with wry amusement but they must have given him pause for thought at the time.

Some of the issues in relation to settlement analysis that were causing difficulty in the 1950s were resolved by a paper that Skem wrote with Laurits Bjerrum. (*Géotechnique*, 7, 1957) Consolidation settlement is shown to result from the dissipation of excess pore pressure, the magnitude of which is a function both of the properties of the clay and of the applied stresses. The relation of the geological history and the properties of a clay to its settlement characteristics are demonstrated. (Burland. Commentary in *Selected Papers on Soil Mechanics*)

Florence

1956 was the year of the eighth 'Congresso Internazionale di Storia delle Scienze,' held at the Villa Favard in Florence. Skem spoke at a session on the

History of Technology and Applied Science on 'The Origin of Iron Beams'. Nancy went with him and they had a wonderful sightseeing holiday in Florence, Fiesole, and Pisa. Skem was particularly impressed by Brunelleschi's 15th-century dome of Florence Cathedral, and Arnolfo di Cambio's 14th-century tower of the Palazzo Vecchio. He noted the angle of the Leaning Tower of Pisa, which, in 1956, was 13 feet out of plumb, and the tilt increasing by nine millimetres a year. Ten years later, he was to be involved in the labyrinthine cogitations of one of the many Italian commissions there have been to consider how to prevent the ultimate collapse of the tower.

The Boltons

1956 was a rewarding year professionally but, on the domestic scene, disaster struck. The peaceful co-existence of the Skempton and Eastwick-Field families at The Little Boltons was suddenly disrupted by John Eastwick-Field announcing that, with his growing family, he now wanted to occupy the whole house and incorporate our upstairs flat as accommodation for them. As John held the freehold, he had every right to do this with sufficient notice. Skem was devastated. He regarded this as a betrayal of trust. He had thought he and John had a 'gentlemen's agreement' but this was turning out to be worthless. Skem depends upon security and stability in his domestic arrangements and any change causes him great anxiety. (In fact, his life is characterized by remarkably little change in his living arrangements. Routines have been established and maintained and deviation from these rare.)

He and Nancy embarked upon house-hunting, a very stressful occupation for Skem. They even, for a time, contemplated moving out to the country and I was very excited to be taken to see a lovely cottage with gabled windows, called *Butchers' Orchard*, near Wendover, which was surrounded by a large garden with fruit trees and fields. (My dream at the time was to own a pony.) I was very disappointed when Skem decided that commuting into London daily was not something he wanted to contemplate. He was used to being able to walk to and from college. Wimbledon was the next idea, and we saw some nice houses there, but that would have involved a long journey on the District line to South Kensington. By this time, Skem was in a state of nervous exhaustion and distress. I remember the shock of seeing him in tears. Katherine and I were hastily bundled off to Scarborough for the Christmas holidays while Nancy and Beatrice rallied round to support Skem, who was sometimes barely coherent and, I think, on the verge of a breakdown.

Fortunately, on January 24th 1957, they saw an advertisement in *The Times* for a flat on the first floor of an impressively spacious stuccoed house a stone's throw away from The Little Boltons, in the more grandly named, 'The Boltons'. (In those days a flat in The Boltons was affordable on an academic's salary.) The flat needed substantial refurbishment and there was a frenzy of discussion of colour schemes, moving of paper furniture round graph paper, on which

Skem had drawn all the rooms to scale, and the designing of carpenter-made kitchen cupboards. I remember trips to Sanderson's in Berners Street to choose wallpaper.

Carlotta Hacker, the newly-arrived secretary, who occasionally came home with Skem to help either him or Nancy with some task or other, reports that the only time she saw Nancy, who was normally perfectly content to play second fiddle to Skem, looking cross with him was when they were moving house. She dropped in to help and there was Skem sitting in an armchair, screwdriver in hand, happily fiddling with a plug and fuse but not really making much effort – holding forth on the marvellous new flat – while Nancy was doing all the heavy work, lugging things back and forth.

The flat had a sunny balcony, where we ate breakfast in the summer, overlooking the oval communal garden and St Mary-the-Boltons Church, with its four black angels round the base of the spire. One of the large rooms became the workroom, shared by Skem and Nancy, with her bookbinding equipment at one end and his two desks and bookshelves at the other. We moved into the flat in June and Skem embarked on a phase of purchasing antique furniture to enhance the new flat and perhaps also with an eye to investment. He bought some beautiful pieces, glass-fronted cabinets to house Nancy's fine bindings and his growing collection of antiquarian books, a sofa

16, The Boltons from The Wood Engravings of Mary Skempton.

table, some console tables, and some Victorian chairs. They were all in keeping with the high Victorian rooms. Nancy hung light yellow velvet curtains at the tall windows and made loose covers from William Morris print fabric. Paintings by her, by John Tunnard, and by Julian Trevelyan and his second wife, Mary Fedden, were arranged on the walls. Skem's own growing collection of historical engineering prints was also hung up. The layout of the flat has remained substantially unchanged for 40 years.

Head of Department

Skem was promoted to Professor of Civil Engineering and thus Head of Department at IC in 1957, in direct succession to Pippard. This meant that he became responsible not only for Soil Mechanics but also for all the other branches of civil engineering – structures, hydraulics, and so on – and for the management and administration of the department. He regarded this last aspect of the job with considerable ambivalence. He rarely called staff meetings, convinced as he was that everyone hated meetings as much as he did and would be only too glad not to have to go to any. This displeased some of the newer members of staff who would have liked an opportunity to hear their own voices in meetings. They wanted him to be more available, to be the 'father of the department'. (Ian Munro, who became Head of Department in 1982 was like this, according to Joyce Brown, Skem's assistant from 1962–72. 'He had an open door at all times, so he was always being bothered by people. He died at an early age of the subsequent stress. Skem often seemed aloof and unaware of people. You could be in the lift with him and he would ignore you, his mind being on the use of dredgers in 1802.') Essential meetings, for example about finance, were never longer than half an hour. Skem's method of problem solving was to sort things out by conversations in the corridor. Staff were never summoned to his office – he would come across someone in the coffee room or go and see them in their room. Things were discussed and agreed in this way and the meeting ratified the agreement. If, by the end of the meeting, anyone was aggrieved, they were simply ignored. When correspondence arrived that he did not want to deal with, he would pick it up by the corner and drop it straight in the bin. This style of management, of course, did not suit everyone but it meant that Skem could keep his focus on what he felt was productive – the research, the consulting jobs, and the teaching. Luckily, the increasing volume of administrative work led to the creation of the post of Assistant Director and R.J. Ashby was appointed.

There were those who objected to Skem's management style, or lack of it. Ian Munro, for example, used to call the Soil Mechanics group 'the characters of the 5th floor'. When, after his retirement, Skem presented his collection of early papers to the library, Munro was dismissive about the idea of his leaving things to posterity. Nick Ambraseys, who joined the staff group in 1958, feels that even the critics later came to regard Skem's tenure of office as a golden period and, when Munro, in his turn, became Head of Department, he often used to say, 'This is an issue we need to resolve in the way Skem would have done.'

By 1957, Skem had accumulated an impressive body of published work and he made the transition to full Member of the Institution of Civil Engineers (MICE) in August of that year.

London Conference

That same year, the Fourth International Conference of Soil Mechanics and Foundation Engineering took place in London. This was a huge event, attended by several hundred people, with Terzaghi as its President. Wynne-Edwards drummed up the funding for it and organized bursaries. Skem was involved in the organizing committee and also presented his paper written with Fred DeLory, one of his Canadian research students, on 'Stability of Natural Slopes in London Clay'. This is another early application of effective stress analysis to slope problems. It does not yet address the issue of residual strength but illustrates Skem's interest in geomorphology. At the end of the conference, Skem was elected President of the International Society of Soil Mechanics and Foundation Engineering, (ISSMFE), in direct succession to Terzaghi.

Terzaghi's presidential address at the ISSMFE London Conference, 1957.

Nancy helped the dynamic Sheila Glossop to organize the programme for the wives of conference delegates, doing 'dry runs' of the planned trips to check journey timings, venues, and so on. (Even today, it is rare in soil mechanics that the conference delegate is a woman and the partner a man.) Sheila remembers a very fierce and determined delegation of Russian ladies demanding to be let into a session at the Civils *after* Terzaghi had begun to speak. This group was not allowed to travel anywhere without their official guide, and went around in their own bus, such were the restrictions of Communism at the time. Ralph Peck brought his daughter, Nancy, with him as a high school graduation present, she joined in the ladies' programme, and father and daughter were invited to dinner at The Boltons on August 20th. Admittedly, Skem was occupied with organizing the conference and was still in the throes of moving house but I am struck by a contrast. Peck had made huge efforts on his behalf when he visited the USA in 1954 (and was to again in 1959) but, for Peck's trip to England, Skem wrote a somewhat offhand postscript sideways in a letter margin, 'Can I help with accommodation?' (letter 11.6.57)

The profits from this conference were used to found the Rankine Lectures, which, from the outset, were envisaged as events of world importance. The lectures are now organized by the British Geotechnical Society and it is they who decide who to invite. They are given alternately by British and overseas lecturers. (They now have such prestige that a recent prospective lecturer took as much as six months' leave of absence in order to prepare for one. Casagrande gave the first Rankine Lecture in 1961 and Skem the fourth in 1965.)

Another event of 1957 was a short consulting visit that Skem made to Turkey. He visited Istanbul and Ankara, and took a photo of a bustling street scene from his bedroom in Bartin. He came back laden with two lovely Turkish rugs and a record of music that sounded very strange to our ears. What made more of an impression on me was the delicious real Turkish delight.

Engineering history

Since his BRS days, Skem had been fascinated by early engineering achievements. His 1950 BBC broadcast, in which he defined civil engineering and soil mechanics (see Introduction), goes on to outline the history of soil mechanics, starting with the earliest developments in the 18th century: 'The French engineer and physicist, Coulomb… published a classic paper in 1776 in which he gave the basic law of the shear strength of soils, a correct theory for the design of retaining walls, and made the first step towards solving the problem of the stability of cuttings, earth dams, and embankments.'

He then describes a second developmental period, culminating in the publication in 1846 of a book by the French civil engineer, Alexandre Collin. This book contains 'the results of a brilliant series of field observations on landslips in clay strata and on the stability of earth dams, and most important of all, a description of an apparatus for measuring the shear strength of clay.' As far back as 1946, he had read a paper on Alexandre Collin to the Newcomen Society for the study of the history of engineering and technology. (It was this lecture on a French engineer that gained him 'brownie points' with M. Caquot in Paris, and Kerisel translated it into French.)

Skem continues: 'By this time (1846)... soil mechanics was beyond doubt a subject in its own right. But a survey of the second half of the 19th century reveals the astonishing fact that there was hardly a single contribution of any lasting significance... It wasn't until the decade 1910–1920 that the subject began to develop again.'

Another area of historical research for Skem was early river navigations. In 1950, he had read L.T.C. (Tom) Rolt's book *The Inland Waterways of England*. Although this was mainly concerned with Rolt's main interest, canals, it also talked about river navigations and made reference to Willan's 1936 book, *River Navigations in England 1600–1750*. This book addressed the economics and political aspects of river navigation but Skem was determined to find out more about the engineers referred to and got in touch with Willan. This was the beginning of Skem's interest in historical biography of engineers, which became a major area of work later in his career. Skem's 1953 Newcomen Society paper ('Engineers of the English River Navigations, 1620–1760') looked at the considerable works which were carried out to improve and extend the river navigations in England prior to the construction of the earliest canals of the Industrial Revolution. The engineers of these works were living in the days before civil engineering had become fully established as a profession and the paper outlines the early developments of the organization of civil engineering, from the work of the early pioneers of the 17th century to the professional consultants, such as John Smeaton, in the mid 18th century. (This paper was later included in the volume *Civil Engineers and Engineering in Britain, 1600–1830* published by Ashgate in 1996.)

That same year, 1953, Charles Singer, a very senior figure, much older than Skem, retired Honorary Professor of the History of Medicine at University College London, was planning a monumental five-volume book, *History of Technology*. The directors of ICI provided funding for this project, which lasted some ten years. Singer knew of Skempton by reputation and approached him to write the chapter in Volume III on 'Canals and River Navigations before 1750'. Rupert Hall from Cambridge was to do a chapter on military technology in the same volume and he took on primary responsibility for editing that volume. (Singer lived for most of the year in Cornwall.) Singer, Hall, and Skem would occasionally meet during the summer months for lunch at the Athenaeum, of which Singer was a member, to discuss Skem's contribution. The book was published in 1957, with very high-quality illustrations. Singer insisted on the illustrations being redrawn for the purpose, rather than copied, which gives a uniform style of presentation.

Iron frames

Skem's interest in engineering history had, by now, become a major preoccupation. In the mid-1950s, he started investigating early iron-framed buildings. Hope Bagenal suggested he contact H.R. (Roy) Johnson of the Architecture Department at Sheffield University, who had done a thesis on the subject. Skem went up to Derby to meet Johnson and look at the iron-framed cotton mills of William Strutt, such as the West Mill at Belper, now demolished, and at Milford, downstream on the River Derwent. Iron frames were an important innovation in mill building, as there was a need to find a more fireproof alternative to the earlier wooden beam structures. Skem liked Johnson very much, 'a cheerful, well-intentioned, and quietly competent man'.

In 1956, Skem and Johnson gave a Newcomen Society paper on Strutt's 'Fireproof and Iron-framed Mills'. Two or three years later, interest in iron-framed mills revived and Skem went north again, this time taking Carlotta with him. Strutt's father had worked with Arkwright, inventor of the water frame for spinning. As well as building iron-framed mills, Strutt had been involved in the setting-up of Derby Hospital. It was in a remote and dusty attic of the hospital that Skem and Carlotta excitedly discovered a damaged portrait of Strutt, which is now restored and hangs in the Council Chamber of the Science Museum in South Kensington. Carlotta nursed fervent ambitions to become a writer, and Skem encouraged her to write her small monograph on Strutt, which was published in 1960 in the *Journal of the Derbyshire Archaeological and Natural History Society*. It contains a photo of the portrait they found. Skem went on to investigate other later mills at Shrewsbury and elsewhere. Work on those came out in a paper written jointly with Johnson in the *Architectural Review* in 1962. (Carlotta went on to write about 20 books, some travel and biography and some for children.)

Skem also took a great interest in the famous boat store at Sheerness Dockyard, on the Isle of Sheppey, Kent. This is an entirely functional, very modern-looking four-storeyed iron-framed building, which had been built in 1861. He visited Sheerness with the distinguished photographer Eric de Mare, who had discovered the building, and their joint article appeared in the *RIBA Journal* in 1961. De Mare took some wonderful photographs of the boat store and Skem addressed the Newcomen Society on the subject, clarifying the historical significance of the building relative to its predecessors and successors among industrial buildings.

Telford Exhibition

July 1957 marked the bicentennial Thomas Telford Exhibition at the Institution of Civil Engineers, a big event in the study of engineering history. Skem worked with the exhibition director, Richard Buckle, who is well-known for his ballet staging but who was also an excellent exhibition designer. They

Eric de Mare's photograph of the Sheerness Boat Store. Courtesy of the RIBA.

jointly wrote the catalogue for the Exhibition, and Skem wrote a companion article on Telford, the 'Master Builder of Bridge and Road' in *The Times*. Skem and Nancy hosted a celebratory lunch party at The Boltons on the exhibition opening day.

L.T.C. Rolt

An adviser to and visitor at the exhibition was Tom Rolt. Tom was a pioneer of the revival of the English canals and, with Charles Hadfield and Robert Aickman, a founder member of the Inland Waterways Association. He had trained as an engineer but had turned to writing. In the summer of 1939, he had toured the canals of the Midlands with his first wife, Angela, in a narrowboat, *Cressy*, and had written a wonderful book about these travels, *Narrow Boat*. In 1957, he was just publishing his famous biography of Isambard Kingdom Brunel and was in the process of writing his book on Thomas Telford. Rolt's book *The Inland Waterways of England*, published in 1950, was a strong

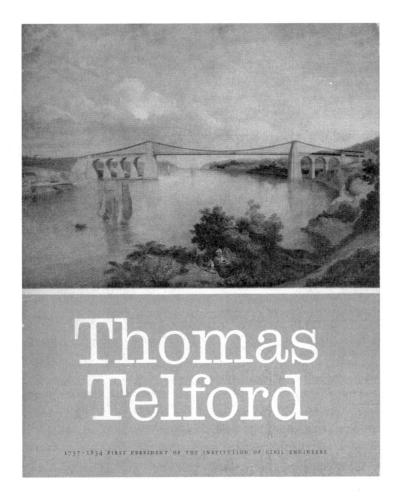

Front cover of the Telford Exhibition catalogue. Courtesy Institution of Civil Engineers.

influence on Skem, who had thought of Tom as a hero for several years. Likewise, Tom was familiar with Skem's river-navigations paper, as Skem had sent him an offprint. Skem met Rolt for the first time in connection with work on this Telford exhibition. Tom's second wife, Sonia (who had previously been married to a narrowboatman) remembers that, when Tom was researching Telford, the Civils was a major source and he frequently travelled from Gloucestershire, where they lived, to London, where he saw Skem. The

Skemptons and the Rolts struck up a close friendship. Soon after the exhibition Skem and Nancy visited Tom and Sonia at their ancient mellow stone cottage in the tiny Gloucestershire village of Stanley Pontlarge. Skem was investigating landslips for the Severn and Trent Water Board on Bredon Hill, where an underground reservoir stored water pumped up from the Avon through a hidden pipeline. Skem and Nancy sometimes stayed at Pershore, the nearest little town, and sometimes at Stanley Pontlarge and did a lot of walking about on the hill. Nancy came back to the cottage with tales of seeing badgers on the east end of Bredon.

Sonia is perceptive about the way in which both Tom and Skem found out about and brought to life the men who dug canals, drained fens, and built railways. They both had an almost empathetic understanding of what it must have been like to create those enormous works. 'Great movings of earth, the whole thing could fail, the poignancy of what a near-run thing these works were, the small man against gigantic odds,' says Sonia. They both had an appreciation of the detail, down to which kind of spade the workmen would have used. Sonia loves the soil-mechanics words, like 'solifluction', which Skem taught her.

While Tom and Skem were smoking their pipes and 'yarning', Nancy would help Sonia in the kitchen. One day they made Nancy's recipe for walnut bread together. Nancy was very knowledgeable about plants and showed Sonia how to take cuttings in the large and beautiful cottage garden.

Rolt went on to write further biographies of Trevithick and the Stephensons, books on transport history and a series of beautifully written autobiographies, *Landscape with Machines*, *Landscape with Canals*, and *Landscape with Figures*.

Department of the History of Science and Technology

In 1953, Skem had introduced an annual series of History of Engineering lectures for students and staff throughout the department at IC. He gave four to seven lectures a term, to which students came voluntarily, as it was not an exam subject. (He continued them until about 1988, when he was aged 74, and, into the 1990s, people still asked if they were still going on.) Mike Chrimes, librarian of the Institute of Civil Engineers, says that the careers of most of Skem's students he knows from the 1950s have encompassed an interest in history because of Skem's lectures.

The Rector of Imperial College in the late 1950s was Patrick Linstead. He was instrumental in setting up the Department of the History of Science and Technology at the college. Skem was a warm supporter of this and Linstead hosted a dinner in August 1959 to discuss the pros and cons of the scheme, to which a great number of people were invited. Charles Singer was called in as an 'adviser'. Skem and Nancy went to stay with the Singers in Cornwall to discuss the project. They found their house, *Kilmarth*, in Par rather grand. It was rented from Daphne du Maurier and had a living room so large that the thousands of Singer's books could be housed on library shelves that stood out at right angles from the wall, as well as lining the walls. There was a fine view over St Austell Bay. Mrs. Singer was a formidably intelligent and very pleasant woman but housekeeping was not one of her strong points. Skem saw her as a pioneer of women's education, 'very Oxford', but Nancy noticed that the kitchen was hung with cobwebs. It was at Par that it was decided to offer the History chair to Rupert Hall, who had edited the *History of Technology*. He was well established in Cambridge and a Fellow of Christ's College but was, at the time, living in the USA. However, in 1963, he responded to the invitation to take up the chair and set up the department. His wife, Marie, joined him. She is a well-known historian in her own right, the unofficial historian of the Royal Society, and the Halls together edited the letters of the first Secretary of the Society, Henry Oldenburg, who lived during the Restoration. Norman Smith was a reader in the department; he continued after Rupert and Marie left and gave the History of Civil Engineering lectures for a short time after Skem retired. (Skem thinks that now the only lectures in Britain on the history of engineering for undergraduates are given by Roland Paxton at Heriot-Watt University in Edinburgh.)

Bookbinding

Rupert Hall has a very large and fine collection of books. At Skem's suggestion, he had 20 or 30 of them either repaired or rebound by Nancy, including a full restoration of a copy of Robert Boyle's works in five folio volumes and a complete cloth rebinding of Lalande's three-volume *Astronomie*. Hall would walk from Imperial College to The Boltons to deliver and pick up books and enjoyed the process of discussion with Nancy in her workroom about how the binding was to be done, and what was in keeping. He would give an opinion as to leather or cloth and she would give her ideas about ornamentation and tooling on the spines and fronts. She had fashioned her own bookbinding tools for lettering and ornaments, which she heated over a little gas stove with a notched ring round the flame to support the turned wooden handles of the tools. She also boiled up a particularly foul-smelling glue in a kind of billycan, which she used to stick book spines. She made special trips to specialist leather merchants in Hitchin and came back with beautiful coloured and textured skins, which she cut to shape with a Stanley knife, for spines and corners or full bindings. She also made bindings for many other colleagues and friends of Skem's and, of course, for his own books. Her charges were very reasonable, Hall thought, and she never made much profit, doing it for the love of the craft.

Although bookbinding was now her main craft, every autumn, Nancy would get out her sketch book, and make a trip to T.N. Lawrence & Son, the wood-engraving block-makers in Bleeding Heart Yard in Holborn (where she was always addressed as Miss Wood, the name in which she had opened an account with the firm in 1936). Skem writes,

Then started the almost miraculous transformation of a plain, mathematically perfect, boxwood block into an object from which with great care a hundred prints were taken; each, if things had gone well, a minor work of art.

At the due time, all Skem and Nancy's friends would receive a woodblock-printed Christmas card.

Rupert Hall liked Nancy very much and found her quiet but very firm, a person not to be trifled with. Although not an engineer himself, Hall had great respect for Skem as an engineering geologist and for his scholarly pre-eminence, as well as strong personal liking. Although he and Marie used to be invited to dinner with the Skemptons at The Boltons, he never felt that Skem was an easy person to know intimately – perhaps very few people knew him on that basis. The Halls remember Nancy always serving roast pork with cabbage cooked with celery, 'Very good it was too; Nancy was efficient in all that she did!' They also heard about the time when Skem and Nancy had to endure Judy Garland as a downstairs neighbour at The Boltons. (This was during her late self-destructive years. She and her much younger husband would party all night and she would rehearse with her daughter Liza Minnelli until the early hours. Next morning, a servant would be sent upstairs to complain when Nancy started up the vacuum cleaner at 10.30 in the morning.)

Foundations

The mid to late 1950s was a period of large-scale construction of office blocks and tower blocks of flats in London and elsewhere. Le Corbusier had led the way in the 1920s with his architectural ideas of the 'Radiant City', composed of skyscrapers within a park. The philosophy of the time was that quality of life would be improved if slums were cleared and bomb damage from the war replaced by blocks of modern flats surrounded by grass and trees. While we may now question the wisdom of this policy, at the time, the social costs of this type of building were not understood, and ideas about the importance of 'defensible space' not yet articulated. (This is a concept popularized by the American architect and town planner, Oscar Newman, who argued that to create areas that could not be overlooked was to invite vandalism and the attendant problems that post-war tower-block developments became prone to.) It was the heyday of town planning, in a pioneering spirit. In order that these buildings could be safely constructed, there clearly needed to be detailed understanding as to how London clay performed under such unprecedented stresses. Skem had written the 'classic' work on foundations and the bearing capacity of clay presented at the 1951 Building Research Congress (see Ch.4). He had developed this work later in the 50s, together with Ralph Peck, Don Macdonald, and with David Henkel. In particular, two important *Géotechnique* papers were published.

The pipe-smoking Skem in 1957, here at King's Lynn. Photograph by Gudrun Bjerrum.

Firstly, in 1957, he, together with Laurits Bjerrum, gave a simple but elegant analysis of the problem of consolidation settlement, which is shown to result from the dissipation of excess pore pressure, the magnitude of which is a function of both the properties of the clay and the applied stresses. Skem and Bjerrum demonstrated the relation of the geological history and properties of a clay to its settlement characteristics. The paper is a model of clarity and still influences the design of foundations in clay. ('A Contribution to the Settlement Analysis of Foundations on Clay')

The second paper was one from 1959, which is Skem's only publication on piles in clay. 'This paper was written at a very opportune time, since piles of this type were being widely used as foundations for the spate of high-rise construction in London and other cities which had just begun. Machines for constructing larger capacity piles were then developing rapidly, and conflicting ideas existed on the mechanism of behaviour and design of these piles.' (*Selected Papers on Soil Mechanics*) The paper analyses a number of records of loading tests on bored piles in relation to the accumulated knowledge of the undrained strength of London clay, and clearly describes the mechanics of the behaviour of the shaft and base of the pile. The paper made a profound impact on consultants and contractors.

Ambraseys and Morgenstern join the staff

1958 brought two important additions to the Soil Mechanics section at Imperial. Nick Ambraseys was appointed to bring engineering seismology into the postgraduate course. In the same year, Norbert Morgenstern, who had come from Canada as an Athlone scholar, and had so distinguished himself that Skem made him research assistant in the Soil Mechanics section (he joined the staff as lecturer in 1960). Skem was particularly friendly with Nordie, as everyone called him, admiring his intellectual brilliance and cultivated and charming personality. He stayed for a further eight years before returning to Canada, where he made the University of Alberta a major centre of soil-mechanics excellence.

From theory to practice in soil mechanics

To celebrate Terzaghi's 75th birthday, Arthur Casagrande conceived the idea of an anniversary volume, and he, together with Bjerrum, Peck, and Skem, formed a *de facto* planning committee with Karl as special consultant. They missed the birthday deadline and the correspondence between them became labyrinthine. What would be the content and who would be the publisher? Particularly controversial was what to include about the influence on Terzaghi of the work of Fillunger on tensile strengths in concrete, which Skem regarded as a 'landmark in the development of ideas on effective stress'. (letter 4.3.59) Terzaghi maintained he had always disagreed with Fillunger. (letter 11.3.59) (There had been a bitter academic dispute between Terzaghi and Fillunger at the Vienna Technische Hochschule in 1937, which had had tragic consequences. Fillunger felt his academic reputation had been undermined and entered into a suicide pact with his wife. They died together.) Skem's diplomatic solution to this dilemma was to write, 'Fillunger's tests fall very nicely into place in their historical context as being one of the series of tests made in the period around 1910–20, which implicitly show the principle of effective stress, but which were not fully understood by their authors'. (letter 22.5.59) This compromise was endorsed by Terzaghi: 'I wish to express once more my appreciation for the splendid job you have performed in unravelling the history of "effective stress".' (letter 3.7.59) The volume finally came out in September 1960 to glowing reviews.

Chapter 6

The 60s – Mangla, The New Building and the Rankine Lecture

Mangla

One of the biggest and most rewarding projects with which Skem has been involved is the Mangla Dam on the river Jhelum in Pakistan. Before India and Pakistan became separate independent states in 1947, the irrigation of the great fertile plains of the Punjab, most of which became West Pakistan, had been developed entirely from the Indus river system. Barrages diverted water into irrigation canals between the main rivers but did not contain the monsoon floods. With partition, the water supply to Pakistan threatened to become precarious, as the headwaters of the Rivers Chenab, Ravi, and Sutlej were on the Indian side of the border and water could therefore be diverted from them for use in India, instead of supplying the irrigated land in Pakistan. The question of water supply was bitterly disputed and was finally resolved by the Indus Waters Treaty of 1960. Under the terms of this treaty, the waters of the three easternmost rivers – the Ravi, the Sutlej, and its tributary, the Beas – were allocated to India, while those of the three western rivers – the Indus, the Jhelum, and the Chenab – were allocated to Pakistan. It thus became important that Pakistan could store some of the monsoon flows in the three western rivers and convey it by a system of link canals to the areas that had previously been irrigated by the eastern rivers. It was also important to generate the hydroelectric power deemed essential for the future of Pakistan.

Mangla, on the Jhelum, and Tarbela, on the Indus, had been identified as the sites for the large storage dams. These, together with new barrages and a network of link canals, comprised the Indus Basin Project. At that time, the complete concept was the largest water development project ever undertaken in the world. To give an idea of the size of the project, apart from the two huge storage dams of Mangla and Tarbela, which were to be considerably larger than Fort Peck in America, then the largest earth dam in the world, some of the linking canals in the network are large enough to carry ten times the average flow of the River Thames at Teddington. The main spillway at Mangla was designed to discharge a maximum flow roughly three times the normal flow over the Niagara Falls.

There were some 2000 engineers and technicians and a labour force of 30,000 engaged on the project for a period of 15 years. The project had international financing and the World Bank administered the fund, the total cost of which was £780 million. (Multiply by 15 or more for current values.)

The Mangla Reservoir is in fact created by three earthfill dams – Mangla itself, Jari Dam and the Sukian Dyke. In 1957, the Government of Pakistan appointed Binnie & Partners as consulting engineers to the Mangla Dam project, in association with the American company, Harza Engineering (for design of the spillway) and Preece Cardew & Ryder (for electrical and mechanical aspects). Geoffrey Binnie, then senior partner of Binnie's, was responsible for the assignment. His reputation as a dam engineer was already well-established and, realizing that the best soil-mechanics expertise available would be essential for the successful engineering of the huge dams at Mangla, he had already retained Skem as the firm's consultant.

Work began immediately on preparing a project planning report which was accepted in December 1958, with what later became known as the West Pakistan Water and Power Development Authority (WAPDA) as the client. The report included proposed outline sections of the dams, based on the site investigations that had been carried out by that time. Skem visited Mangla for the first time in April 1958. Alan Little, Binnie's chief soil-mechanics engineer on the project, was anxious to have his advice on the strength parameters that should be used for the preliminary design of the dams.

Prior to Skem's first visit, small-strain triaxial tests had been done on some borehole samples of the clay stone and some basic calculations made about factors of safety. In those early days, Skem devoted much study and time to developing a thorough understanding of the geology of the reservoir area and its surroundings. He was very interested in the earlier work done by geologists of the old Indian Geological Service. As events were to prove, he showed great foresight in developing this line of enquiry. As Geoffrey Binnie points out in his paper, 'Engineering at Mangla' (reprinted by Binnie's in what is now known as the blue 'Mangla Book'), the geological history of the site has had a more profound influence on the design of this project than on most other comparable jobs.

The project is located in the low hills southwest of the Himalayas and the geology is composed of beds of the Siwalik system. These are freshwater sediments deposited in the Plio-Pleistocene age, made up of layers of sandstone, siltstone, and clay, overlain in places with more recent gravel. The deposits are, in soil-mechanics terms, heavily overconsolidated, owing to erosion after their deposition. The Siwalik formation has been subjected to folding and faulting during a period of uplift of the Himalayas. One of the major thrust faults at the foot of the Himalayas passes within 30 miles of the site. The area is, therefore, seismically unstable and destructive earthquakes have occurred. The clay beds presented major problems to the design of all parts of the project.

Between 1959 and 1961, while contract designs were being developed, Skem made several visits to Mangla, spending between six and ten days at the site, as well as being constantly consulted by Alan Little and the staff at Binnie's London office. At Mangla, he worked, in particular, with Richard Phillips, Binnie's resident engineer. Resident engineers on the Indian subcontinent at that time were often of the expatriate Indian Army type. Phillips had a distinguished war record, which, in his gentlemanly way, he never referred to. Phillips returned to London at the end of the pre-contract site-investigation stage. Skem remembers that he was a keen cricketer. Skem loves watching cricket and he met Phillips many years later, when he was playing for a team rejoicing in the title, 'The Stragglers of Asia' against the home team at Hurlingham Club.

On early visits, Skem stayed in a Pakistani-style house built for expatriate staff on the old Mangla colony on the Left Bank. On his way to the site from Lahore, he took several photos of village life and farmers in turbans. On later visits, during the construction phase, he stayed in one of the western-style bungalows with air conditioning, built specially for the Europeans. Each bungalow had its own staff. The head servant would always wear immaculate white robes and a turban. Skem remembers a social gathering on one early visit at the Left Bank colony, where chairs were ranged round the perimeter of the room and engineers and their wives made formal, somewhat stilted conversation across the large space in the middle, and 'rather strange' food was served. (Skem has conservative tastes and would regard anything containing curry as 'strange'.) The older men wore Pakistani clothes but the younger engineers were thoroughly Westernized and most of them had been to university in the UK or the USA.

Samples of the Mangla materials were initially taken to London for testing at Imperial College but, in 1960, a soils laboratory equipped with apparatus supplied by Soil Mechanics Ltd was set up on site. Subsequent to that, all analyses were made using effective stress values. The contract designs of 1961 were based on what was then considered to be the 'ultimate' strength.

There were four tenders for the contract, which were opened in a public event at Caxton Hall, Westminster, in November 1961. Gordon Eldridge, a member of Binnie's team on Mangla, remembers the excitement of the occasion, with the bid envelopes being opened one by one. Binnie's won the contract and, in January 1962, construction work began in earnest. Skem made the first of several visits during the construction stage in November 1962. Binnie's site-supervision staff were all quite newly arrived at that time and Gordon Eldridge remembers Alan Little announcing the arrival of Professor Skempton, warning everyone how eminent he was and how he must be treated with kid gloves and, at all costs, not be allowed to exhaust himself. This warning was entirely superfluous. Skem set himself a demanding schedule, never appearing to suffer from the heat (around 105° Fahrenheit) or disease, to which a number of Binnie's men succumbed. He would set off on site investigations early in the morning to avoid the heat of the day, and his stamina and enthusiasm taxed to the ultimate the endurance of the men accompanying him.

During this visit, Eldridge's wife, Barbara, was called upon to provide a buffet party for Skempton and about 50 guests. She had great difficulty locating suitable English ingredients and spent two whole days cooking and preparing. With his lack of awareness of domestic matters, it is doubtful that Skem appreciated the amount of effort that had gone into this meal.

In 1961, shear zones had been first discovered at Jari and, between then and October 1962, several more were found there. A paper by Binnie and John Clark tells that the site staff nicknamed them 'chapati zones' after the shape of the local unleavened bread. At that time it was thought that the shear zones were confined to Jari, where the dip of the bedding is steep but, as work proceeded, they were subsequently found in the clay beds at Mangla also and, finally, in 1964 at the Sukian site as well. They extended over a large area and were thin fissured zones within the clay beds that were parallel to the bedding planes. Some zones were formed of hundreds of such closely spaced planes that, instead of 'chapati', these were aptly named 'millefeuilles'.

Skem, together with Peter Fookes, Binnie's resident engineering geologist (and later staff member at IC), realized the significance of these newly-discovered shear zones, which were central to how the material behaved. He later wrote, 'During a site visit to Jari in April 1961 I first realized that in the shear zones... the residual strength alone could be relied upon. But at that time little was known about residual strength or its measurement.' (Mangla book p. 271) Clearly, the strength of the shear zones along which the clay had fractured was less than in the adjacent intact unsheared clay. This lower strength would apply to any potential slip running parallel to the bedding, a matter of consequence, especially at locations such as the Mangla intake, where the dip of the strata is unfavourable to stability.

As a result of these fateful findings, it was essential to measure accurately the true residual strength of the clay. Binnie's engineers started a second campaign of sampling and triaxial testing in 1963. These samples were taken mostly from trenches, as opposed to borings. The lower strengths, compared with those originally used in 1961, necessitated a major redesign of the Mangla Dam and the spillway. The new tests showed that the strain to which samples had previously been taken was much less than was required to reduce the strength of the clay to the residual. These findings were very controversial. The other firms involved in the project, Sir Alexander Gibb for the World Bank, and Harza Engineering, needed some convincing about the need for the redesign. Dr Willis, Harza's geologist, for one, did not accept it. He is reported as going round the site with a geological hammer, hitting vertical clay surfaces, saying they had stood for hundreds of years and were not going to fail. Because the area is arid, the clay had the appearance of being rock-hard. (Put it in water,

Mangla Dam, showing the main dam to the left of the Fort Hill and the intake embankment to the right. In the foreground is the River Jhelum and the bridge. Courtesy of Binnie, Black & Veatch.

though, and it will soften. Appearances can be deceptive.)

The consequences of the failure of such a large dam would be unimaginable but the cost implications of the proposed changes were enormous in both money and time. It was suggested that an advisory panel be set up to consider the implications of the findings and present the proposed change of plan to the client, Water and Power Development Authority, (WAPDA). Skem made a further visit to Mangla to attend the advisory panel and his participation was crucial. All-day meetings were held and Skem took the panel members round the shear zones, showed them the test results, and conducted them round the labs. The day was eventually won and the advisory panel endorsed the principles of the redesign in November 1964.

Walton's Wood

Skem's work was a significant contribution to the factors included in the redesign, but, as events would show, the residual strength was *still* being incorrectly measured. It was during the late 50s and early 60s that Skem was developing his ideas on slope stability and residual strength. He was studying this, for example, at Walton's Wood on the M6 motorway in Staffordshire between September 1962 and December 1963. He was called in by Soil Mechanics Ltd, who were investigating a failure there, and it was Silas Glossop who first spotted that this was an old slip. It had been sliding for about ten thousand years.

This research showed, for the first time, that the clay strength on the actual shear surfaces themselves was much lower than that in the body of a shear zone. These findings were later summarized in Skem's Rankine Lecture, published in June 1964. In retrospect, it seems obvious that only the true residual strength of the clay can be relied upon and it is difficult now to imagine how controversial this was in 1963. Skem remembers arguing at length with Peter Fookes at Mangla to try to convince him of the differences. The Walton's Wood findings introduced an entirely new element for Mangla and it was clear that yet another campaign of sampling and testing needed to be embarked upon. Skem made detailed drawings of shear zones during his visit to Mangla in November 1965.

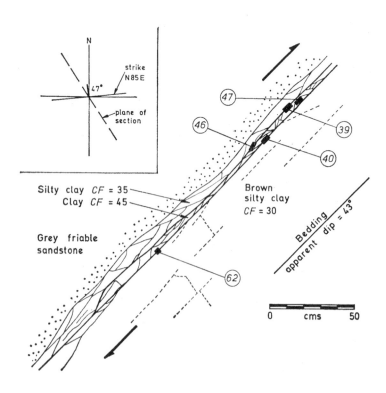

Shear zones D at Jari showing the principal slip surfaces and the countless shear lenses.

Binnie's Mangla report goes on, 'In 1964 the shear boxes at Mangla (supplied by Soil Mechanics Ltd) were adapted to allow the reversal shear box technique.' Skem says that the first reliable shear-surface tests had been carried out by the end of 1964 and the testing continued throughout the following year. Yet more shear zones were discovered during this period, partly because construction work exposed new zones. By then, Binnie's had several engineering geologists working flat out cutting trenches to get material for testing. Residual strength was at last being correctly measured on shear surfaces and in the new reversal shear-box testing techniques. Because shear zones had been found in two-thirds of the clay beds sampled, probably extending throughout their length, Skem advised that they must be assumed in design to exist in any clay bed and be continuous throughout its full extent.

A residual phi value of 18° at Mangla and 13° at Jari was derived. At Jari, no redesign was needed because more sandstone was found than was expected but a yet further, and final, redesign was needed for the Mangla Dam, spillway, and intake, and at Sukian. These were completed in 1965 and the project was inaugurated by the President of Pakistan in 1967. The project has performed faultlessly ever since. Skem's work on the soil-mechanics aspects of Mangla was a major contribution to its success. The possibility of raising the dams is now being studied.

Nick Ambraseys, who was with Skem at Mangla as an expert on earthquakes, describes the way Skem worked at Mangla and other consulting projects. Skem only took on projects that interested him. He could sometimes forget the problem for which he had been called in if a principle or question mark arose that he would want to pursue. For example, on one site visit at Mangla, Skem and Ambraseys saw some fissures in shales. In the car on the way back, Skem was quiet and then, over a whisky at the bungalow, he said, 'Nick, what do you think about that thing we saw today? It's extraordinary. Why was it there? We'll have to look tomorrow at that other place.' And that would start him onto something that occupied him for another three or four years, getting him into sedimentology, and other areas he had not considered. He would pull together information from a number of sources, write to people, and carry out research. Ambraseys continues: 'You have to fall in love with the big projects and work day and night and the projects were part of our research. We would throw in a couple of extra tests which were not strictly essential for the project but were very interesting for us. Binnie's would expect that and they had no objection.' The process was one of taking small steps, asking questions, putting things together, and trying to understand. The explanation would be a practical common-sense explanation, not esoteric, and *never* done on computer, of which Skem had and still has no understanding. He would not accept the computer analyses of slope stability done by Binnie's staff without also being given an analysis of the same problem done by hand and which gave a similar result. A slide rule was the only aid to calculation that Skem ever used. 'On a slide rule, two times two equals approximately four – you are not seduced by absolute accuracy,' he says. Developments in computer analysis and forecasting have now completely overtaken him.

Roger Brown, Alan Little's assistant in the early 1960s, remembers Skem's ability to keep his attention on 'first-order' effects and how annoyed he got when people at meetings were distracted by the very many and complex second-order issues.

Skem had a respect and liking for Geoffrey Binnie but there were some difficulties. Roger Brown tells of one occasion when Skem announced to Binnie that the dam, as planned, could not be built at Jari and then promptly disappeared off for a holiday to Venice, much to Binnie's annoyance. According to Ambraseys, Binnie (unlike most senior engineers) ran the consulting group rather like an army platoon and tended to issue orders. On one occasion, for example, he took some decision on the group's behalf, consulted a government minister, and sent a copy of a letter to Skem. This was one occasion when Skem got really angry, and Ambraseys remembers him using the dreaded sentence, 'This is uncalled-for.'

Despite these upsets, Skem and Binnie developed a close friendship. One thing that they had in common was that they were both excellent listeners. Anyone, however junior, who believed that he had something important to say, could be sure that both Skem and Binnie would listen carefully and, if convinced, would accept a well-argued and well-founded case. They were both absolutely dedicated to making a success of Mangla. Roger Brown recalls Binnie saying to him, 'We could never have built this dam if the staff had not all been *gentlemen*.' Binnie clearly included Skem in this category.

There is now an awareness of some of the disadvantages of large dam projects. There is the problem of sedimentation, which reduces the capacity of reservoirs. There are undoubtedly ecological implications, and the politics around displacement of populations are often unsavoury. Everyone I have talked to about the Mangla project feels that it has brought enormous benefits to Pakistan. Skem speaks passionately about it. The area was formerly sparsely populated, the villagers had a precarious existence as subsistence farmers, and flood and drought were commonplace. The villages that were built to rehouse the displaced population, Skem says, are of a much higher standard than their previous homes. The town of Mirpur was moved to higher ground. The reservoir of the Jari Dam was in Kashmir, whose inhabitants had retained their British passports at Partition. Many of them took the opportunity of having to relocate to move as far as Northern England, using their precious passports. Large numbers settled and found work in Bradford. Those who made the move spread the word to their friends and relations and, soon, BOAC set up business in an office in New Mirpur. Many have now retired and returned to the dam region in Kashmir. The dam has brought electricity, irrigation, and fertility to an area much threatened by the Indian need to harness the headwaters of

Pakistani rivers. S.S. Kirmani, Chief Engineer for WAPDA, told Gordon Eldridge in 1981 that Mangla had paid for itself, in terms of agricultural benefits and power revenues, within two years of its completion. The people value the impressive 453-ft.-high dam, and maintain it well. It has become something of a tourist attraction. On his most recent visit, Eldridge came across a busload of school children at the dam who had been brought on a visit.

While he was involved in the geological investigations in Pakistan, Skem discovered a locality, south of Jari, that contained the fossilized remains of mammalian fauna. His IC colleague John Hutchinson remembers him in his office, 'writing out lists of diagnostic mammalian fossils for the main glacials and interglacials, an interest stemming from his encounter (during work at Mangla) with the famous Pliocene and Early Pleistocene elephants of the Siwaliks of the sub-Himalya.'

As well as their usefulness in helping to date the strata, Skem became fascinated by these fossils and took the best specimens to the Natural History Museum in South Kensington. John Clark, another Binnie's engineer, remembers Skem's excitement about them and reports that the museum subsequently organized a series of visits to the site over several years and have found the fossils of several previously unknown animals.

Northern Ireland and music

To go back to 1958, Skem and Nancy made several much more minor journeys. In the September, they travelled with Ove Arup to Northern Ireland for a Civils conference. They visited the Mountains of Mourne and Carlingford Lough, Stewartstown, the Round Tower at Antrim, and the Giant's Causeway. Skem took a particular interest in the Newry Canal.

That same September, Ralph Vaughan Williams died in his sleep at the age of 86 and was buried at Westminster Abbey. Skem attended the commemoration and funeral service and mourned the passing of a man he regarded as a musical hero.

Hurlingham Club

In March 1959, Skem and Nancy decided to join the fashionable and expensive Hurlingham Club, which they had heard about from a neighbour in The Boltons, and took up the elegant but deadly game of croquet. Mallets appeared in the hall at The Boltons and competitions were entered into. Nancy made a sheet of cardboard into the shape of a croquet lawn, and she and Skem would study the rule book, moving coloured tiddlywinks round the cardboard to work out what the book was telling them (Skem was undertaking a new activity with his usual thoroughness). New friends were made, many of an ex-army type, quite different from their normal circle of academics, practising engineers, and artists. Hurlingham is on the north side of the Thames, in Fulham, in a beautiful classical white-colonnaded stately house, surrounded by gardens, tennis courts, and a cricket pitch. It had included the famous polo club on what is now the adjoining sports field. The fact that Skem and Nancy joined shows that their financial position had improved considerably since the early days in London. Katherine and I played tennis and swam in the open-air pool while our parents laboured to improve their croquet handicaps on the immaculately manicured lawns with the striped canvas pavilions. They met with some considerable success. The *Daily Telegraph* of August 19th 1961 announces in small print at the bottom of the sports page, in a column entitled 'Hurlingham Croquet', that the Longworth Cup was won by A.W. Skempton and the Ladies' Field Candlesticks by Miss E.J. Warwick and Mrs A.W. Skempton. Skem beat General Wilson-Haffenden in the semi-final of the handicap singles but was defeated in the final. Later, Nancy became an enthusiastic member of a ladies' group at Hurlingham called 'Finance for Fun'. They built up imaginary share portfolios and followed their fortunes on the stock market. Another keen member of this group was Eileen Donald, mother of a friend of Katherine's at St Paul's. Eileen's husband, John, was Hurlingham's auditor. They lived in Rivermead Court, overlooking the gates of Hurlingham.

Back to the spring of 1959, and Skem and Nancy made a trip to the Cotswolds and stayed for the first time at the beautiful 16th-century pub, which became a favourite haunt, the Lamb at Burford. Then, after the usual Scarborough holiday and annual trip to Thornton-le-Dale village agricultural show, Skem undertook a second US trip to lecture at MIT. He went to see Ralph Peck at Urbana, who had again been busy responding to Skem's requests for support of various kinds. Peck had suggested a one-month appointment at the Talbot Laboratory but Skem decided against this, as he was so busy with

With Nancy on holiday at Hickling Pasture.

the design of a new building for the Civil Engineering Department at IC and a backlog of unwritten papers. He did revisit Chicago, where he was thrilled to be able to view microfilm, in the City Library, of the original drawings of 19th-century skyscrapers in downtown Chicago. He photographed the Marquette Building, built in 1895.

The Athenaeum

1960 started auspiciously, with a notification to Skem from the Secretary of the Athenaeum Club in Pall Mall that he had been elected a member. Skem felt much honoured but, unlike Terzaghi, who revelled in the atmosphere of the London gentlemen's clubs, he has never spent a great deal of time there dining or hobnobbing with bishops or, indeed, snoozing in leather armchairs, although he was most impressed with the library. He invited the whole family to a lunch in the Garden Room on 6th June 2000 to celebrate his 86th birthday and his knighthood, which he received from the Queen that day.

In May 1960, there was another lecture tour, this time to Madrid. Nancy went too and they visited Toledo and Segovia, with its impressive Roman aqueduct, and stayed with Spanish engineer José Sallas and his wife, Maria.

Studio portrait in the 1960s, by Walter Bird.

That autumn I went up to Birmingham University to read English. Skem had recommended that particular course to me and I felt may have 'pulled strings' in the background, using his connection with Kolbuszewski, because I was given an easy ride in the selection interview. The red-brick Victorian buildings of the university were surrounded by a building site, as it did its 1960s expansion but the English Department was graced by such teachers as Richard Hoggart, Malcolm Bradbury, and David Lodge. Skem helped me move my belongings into the hall of residence and he was also interested in the Barber Institute of Fine Art, just down the road in Edgbaston.

FRS

On March 16th 1961, a telegram arrived for Professor Skempton at the Imperial College of Science and Technology, Prince Consort Road: 'WE HAVE THE HONOUR TO INFORM YOU OF YOUR ELECTION TODAY AS FELLOW OF THE ROYAL SOCIETY.' Skem was only 46 years old. He had been put forward for election several times previously, probably nominated by Professor Pippard (who had kept in close touch with the department after his retirement). The process of election is that about 20 names are submitted and considered by a sectional committee, who put a short list of three forward to the Council of the Society, along with the nominations of the other sections. Whether you are elected depends not only on your own merits but on those of your competitors. (Bishop was high up on the list on several occasions but was never chosen. Ambraseys later went through this process, only to be knocked down at the Council stage, to Skem's distress.) Skem is still only one of a handful of civil engineers who are Fellows. Skem wrote of his feelings to Terzaghi in a moving hand-written letter:

> This is a great honour for soil mechanics in England, and I have been the lucky recipient of a distinction which, by its very nature, cannot be at all widely distributed. Of course, you know that I owe almost everything to you, but this occasion gives me the opportunity of saying so in the most definite terms. Without you, Karl, none of us would have done anything really worthwhile in soil mechanics. Nancy joins me in sending our love to you and Ruth.

In later years, when further honours were heaped upon him, Skem said to me that the FRS meant more to him, perhaps even than the knighthood, which he finally gained very late in life.

The summer of 1961 was packed with conferences for Skem and Nancy. In June they were in Rome at the 7th Congress of the International Commission on Large Dams. They stayed at the Grand Hotel de la Ville, and Skem's album contains glossy invitations to, among others, a reception with the British Ambassador and a dinner at the Villa D'Este. They moved on to Venice, where a photo was taken of Skem and Nancy looking glum on a nocturnal gondola ride.

Paris Conference

These festivities were but a lead up to the really big event of that July, the 5th International Conference of Soil Mechanics and Foundation Engineering in Paris. Skem was chair of the conference in his capacity as President of the ISSMFE. In preparation for his presidential address he took private French lessons. A lady came to the flat twice a week for many weeks. This was, apparently, of only mixed efficacy. Skem does not have a gift for languages. According to Ambraseys, at the end of an after-dinner speech, giving permission to smoke, instead of *fumer* meaning 'to smoke', Skem used the word *fumier*, meaning 'manure', remaining blissfully unaware of his *faux pas*.

At the conference, Skem used his presidential address to sum up the progress of the profession since the first international conference at Harvard in 1936. Peck sums up Skem's speech: 'First, he considered it proved beyond all doubt that "the engineering behaviour of soils in practice can be analysed in a rational manner". Second, he noted the wide acceptance of soil mechanics by the civil engineering profession and by the universities. At the same time, he could not help expressing the concern that young engineers fresh from these universities would, in the possession of their new knowledge, be tempted to carry out designs without developing an intimate knowledge of the sites and an appropriate understanding of soils and rocks.' Peck felt as strongly as Skem about the importance of observation in the field.

In addition to his presidential duties, Skem gave a paper on horizontal stresses in an overconsolidated Eocene clay. He used the data provided by a series of carefully executed oedometer and triaxial tests on London clay from the site of the Bradwell nuclear power station in Essex. John Burland comments that, at the time, his conclusions, that the ratio of horizontal to vertical effective stresses in undisturbed clay could be so high, must have been surprising. The paper attracted a complimentary discussion from Terzaghi, and subsequently, others have measured similar high values in London clay, confirming Skem's 'bold conclusions.'

Terzaghi could not come to the Paris conference because he had suffered an aortic aneurysm and had had to have a leg amputated. But a new generation was emerging. It was at this conference that the young John Burland, later to succeed Skem as Professor of Soil Mechanics, first encountered Skem. 'Skem was the distinguished president. I was in awe,' says Burland. Skem had advocated that the modest Casagrande should succeed him as president, writing to Terzaghi, 'No-one else is so suitable and there can be no possible reason why he should not accept it this time, as the conference will be in North America.' (letter 6.4.61)

After the Paris conference, Skem and Nancy travelled as far south as Avignon, with various colleagues, including the Sallases, whom they had visited in Madrid the previous year.

Cambridge University

Skem gave a series of lectures that year, at Cambridge University School of Architecture, on the history of structures from Roman times to steel structures. Nancy accompanied him for this first one, in April 1962, and they stayed with the irascible Ken Roscoe and his sociable wife, Janet, in their imposing Edwardian house on Millington Road. Subsequent entries in Janet's visitors' book show that Skem came again by himself in May 1962, 1964, and 1965. Janet is an excellent cook and a committed hostess. She was anxious to create enjoyable social events to smooth over the difficulties sometimes created by Ken's abrasiveness. Skem liked and admired Janet very much, finding her 'very civilized and fiercely protective of the reputation of her husband'.

John Burland tells of the rivalry and animosity that existed between Soil Mechanics sections, at Imperial College under Skem, and at Cambridge under Roscoe. This was on two dimensions, firstly personal, particularly the difficult relationship between Roscoe and Alan Bishop. For Roscoe, who had spent so long suffering from his war service and as a prisoner of war of the Germans, the fact that Bishop was a pacifist must have been hard to bear. Worse, Bishop reminded him of his wartime prison interrogator. The other dimension was the different approaches to soil mechanics of the two institutions. Cambridge had developed what is called the 'critical state theory', which they thought was unimpeachable. It was very elegant mathematically but Bishop demonstrated that it did not stand up when applied to real soil and clay. While Cambridge did consulting, mainly with the Government on basic principles, Imperial College was involved in big projects with consulting firms such as Binnie's and Ove Arup. The IC solutions to soil-mechanics problems were worked out in response to real situations on the ground.

A little memorial volume compiled by colleagues after Roscoe's untimely death in a road crash gives insight into his personality, which perhaps explains some of the difficulties. 'He was a rebel at heart and, as such, needed always to ascend so as to be able to implement his will. It was this spirit of rebellion in him which inspired his many attempts to escape from wartime POW camps and later helped to fire his ambition to achieve both recognition and status for himself and his research group,' wrote Alexander Barker. 'Nerves and sensibility were close to the surface. There were no shades of grey on any issue. One result of his nervous disposition was that the same amount of effort was always applied to both big and small problems. In personal relationships, this was often disastrous, since mountains were made out of molehills,' wrote Dicky Bassett. 'Success with Roscoe meant a complete commitment of one's whole personality to his aims,' wrote John Burland (who worked under Roscoe's supervision while doing his PhD at Cambridge). His staff and students called Roscoe 'The Boss', and all were agreed that he had the most enormous capacity for hard work.

His successor at Cambridge was Andrew Schofield, who Skem feels was a loyal supporter of the Roscoe approach. Schofield came from Manchester University and there was also animosity between that school, where the Head of Department was Professor Rowe, and Imperial College. Rowe's department was smaller than those at Cambridge and Imperial College, and he rather resented Skem's influence at the Royal Society. The perception was that IC was the centre of a spider's web on influence in powerful places in London.

Rowe was particularly prone to personal identification with his work. Nevertheless, Burland feels that it is a weakness of Skem's that the destructive tension between IC and Cambridge and Manchester continued. Perhaps the critical state theory may have gained a wider currency internationally were it not for the antagonism of IC. On the other hand, competition between centres meant that people went back and sharpened up their thoughts so, in that sense, the competition was constructive. Skem himself got on well personally with both Roscoe and Rowe. Relationships between Cambridge and IC should shortly improve, Skem anticipates, as Schofield has recently retired from the Cambridge chair, and has been replaced by Robert Mair, who, although a Cambridge PhD, has been head of the Geotechnical Consulting Group (GCG), a smallish but influential consulting group, where Peter Vaughan is a director. GCG and IC work together closely and Mair shares the Imperial College outlook.

Newcomen Society

In December 1962, Skem gave a paper entitled 'Portland Cement 1843–87' to the Newcomen Society meeting at the Science Museum lecture room. In the course of some other work, he had become interested in the material and found, to his irritation, that little was known of its history, so he decided to investigate it himself. Similarly, when he needed to know about dredging, he found nothing could be found about the development of dredgers, 'so, heigh ho, I'd have to do the work myself', and, in 1975, he gave a one-off paper to the Newcomen called 'A History of the Steam Dredger'. The Newcomen had started way back in the 1920s and existed to spread knowledge of aspects of the history of all branches of technology. It provides an outlet for just this kind of idiosyncratic studies and is full of 'eccentric types' with many and varied interests. Annual summer trips are organized. (See Chapter 7)

Civils Library débâcle

While the first cars were driving along the newly completed M1 motorway, Benjamin Britten's *War Requiem* was given its first performance at the rebuilt Coventry Cathedral and President John F. Kennedy was battling with the Cuban Missile Crisis, Skem faced a crisis of his own.

The Institution of Civil Engineers was contemplating building works. The Secretary at the time, Alexander Mcdonald, who was 'slightly mad', according to Skem, wanted to make a restaurant in the basement and persuaded the library committee to dispose of books stored there in stacks to make way for it. There were priceless antiquarian books among the collection. They had always been available in the library by request.

Richardson, the librarian at the time, was unable to stand up to Mcdonald. There was a small announcement in the news pages of the *ICE Proceedings* inviting people to collect anything they might want but Skem had missed seeing this. Richardson called him in desperation and he rushed to the Civils and discovered to his horror a group of workmen taking packs of books, as many as they could hold in their two hands, from the shelves, stuffing them into sacks, passing them up through the basement window, and tipping them into skips. They were to be taken for pulping. Because of his historical research, Skem was more aware than perhaps anyone else of the quality and uniqueness of the Civils book collection. He immediately phoned Glossop at Mowlem's and colleagues at IC, telling them to come immediately in taxis. David Henkel set off in his car in hot pursuit of the first lorryloads but they reached Bowater's plant on the Isle of Sheppey before him and immediately tipped the loads into the shredder. Meanwhile, Mowlem's provided vans and a gang of five or six people from college, plus library staff, including Marjorie Carter, then a junior assistant librarian, worked for days filling vans and taking the books to IC. Everyone got filthy. Marjorie was in overalls and remembers going up or down in a lift with Skem, who was very grim-faced. She said, 'I think this is absolutely appalling,' and he said, 'So do I.' Poor Richardson suffered a breakdown because he had been unable to save the collection.

Mike Chrimes, the present librarian, waxes eloquent about the monstrosity of just pulping 18th-century books. He arrived 25 years later, to find those books that Richardson had managed to save stacked in cupboards from floor to ceiling, where they had just been stuck among pipes and conduits to hide them. Then nothing further had been done. Mike has been able to analyse what was saved and what was lost. Most of the 19th-century material down to 1915 was listed in printed catalogues, so, as he went through recataloguing for the purpose of computerization of the library, he made a note of the material which had disappeared. This was a depressing experience but at least he now has a record of what they haven't got. If it weren't for Skem's knowledge of the collection, much more would have been lost. In his early historical writing, Skem had referenced works that now no longer exist.

Mike finds it frustrating that, for example, in connection with refurbishment work, people come into the library and ask a quite basic question about how iron beams had been designed. He could point them to Skem's work on Sheerness Dockyard but some of his sources, for example, Tredgold, which appeared in the old printed catalogue, had been destroyed. (The first, second, and fourth edition of his *Strength of Cast Iron* has fortunately survived and is in the Civil Engineering library.) Mike is doing his best gradually to replace some of the most important publications.

Meanwhile, at IC, Skem gave some of the more general books to the college library, now housed in a modern building that also contains the Science Museum library. Not long after the dramatic book removal, Marjorie Carter was astonished to return from lunch one day to have Skem come up to her as she was hanging up her coat and say, 'I was just wondering if you would come to Imperial College and be my librarian?' She answered, 'This is very unethical, Professor.' 'Yes, I know,' he said, 'But perhaps you would speak to Mr Richardson?' The department at IC didn't have a library, as far as she was aware. She, nevertheless, went to Richardson and said, 'I've just had this extraordinary offer from Professor Skempton!'

'Take this chance,' said Richardson.

When she arrived at IC, Ashby proudly showed her a tiny room, in which books from the Civils were piled up. She had no job description, as such, but was made aware that, as the college was expanding, the Civil Engineering Department was going to have its own library and she was expected to set it up.

The new building

Following the Robbins report, the early 60s was a period of massive expansion of the old universities and building of six new ones on greenfield sites. Student numbers at Imperial College had grown since the war from 1100 to nearly 1700, causing pressure on space and inhibiting the development of some branches of research. The University Grants Committee had considerable funds to disburse for building development. The Imperial Institute (where Skem had done his Highers practical exams) was to be demolished, to make room for the Civil Engineering and Electrical Engineering buildings. The Rector, Sir Patrick Linstead, asked whether, when the Institute was demolished, it was possible to leave the striking green copper-domed Collcutt Tower standing. Skem made investigations into the bearing capacity of the clay and the settlement of the foundations. Borings were done by Soil Mechanics Ltd. Questions were asked in the House of Commons. It was a nail-biting decision. In a report written with the architects, Norman and Dawbarn, in July 1957, Skem concluded that it would be possible for the tower to be left free-standing but the foundations would have to be extended and the lower part of the tower strengthened. It then transpired that the bells in the tower were the property of the Queen. The outcome was that the tower (now known as the Queen's Tower) stands alone like a campanile, surrounded by a garden in the middle of the campus.

In all the work around the tower and the new buildings, the necessity for close consultation between Skem and the Rector led to a mutually respectful friendship between the two men. Lady Linstead commissioned Nancy to make a wood engraving of the rector's house, a beautiful 1888 Norman Shaw building at 170 Queens Gate, for her Christmas card of 1964.

Skem and the Rector were called upon to entertain royalty at the new

The new Civil Engineering Building, from A Hundred Years of Civil Engineering at
South Kensington.

building. The Duke of Edinburgh paid an 'informal visit' on 12th July 1962 to
unveil a tablet to commemorate the new Engineering buildings, to be followed
by his daughter, the young Princess Margaret who performed the official
opening on 8th October 1963. A photograph of the occasion shows Skem
resplendent in his academic robes, and Sir Patrick looking inscrutable.

In regard to the new buildings, 'the college gave each department a fairly
free hand to make its own proposals within the restrictions on size which were
mutually agreed. Greville Rhodes of Norman and Dawbarn had many
meetings with Skempton and Ashby to formulate the requirements of the
Department.' The UGC allowed so many square feet for different members of
the staff hierarchy. 'An easier way to express this was to count the number of

Princess Margaret visiting the new Building, from A Hundred Years of Civil Engineering at South Kensington. *Left to right: Skem, Prof. Sparkes, Princess Margaret, Lord Snowdon, Sir Owen Saunders, Sir Patrick Linstead.*

windows as a measure of one's status; four for a professor, three for a reader and two for lecturers and secretaries!' (Brown). Soil Mechanics was to occupy the fifth floor. The entrance hall and strategic points on staircases and landings of the new building were embellished by beautiful photographs of bridges (and of the Sheerness Boat Store), commissioned from Eric de Mare. Joyce Brown writes, 'Our move into the new building in April 1963 is still remembered... Some academic staff never recovered from the disturbance of the strata of papers and books in their rooms in the old building but the move provided an admirable opportunity to leave unanswered, for ever, certain academic correspondence in the pending tray!'

Under the old régime, members of staff had taken any textbooks they needed into their rooms and kept them there. One of the first things that Marjorie Carter did on taking up her post as departmental librarian in 1963 was to circulate staff and research students asking them to kindly return these books. She had her work cut out educating the department as to what a library was. After some reluctance ('But I've *always* had this,'), she was inundated with the contents of people's offices, things they didn't want any more. Runs of periodicals, American journals, came to light, but the books she wanted returned tended to remain in rooms, especially Skem's. The library staff were trained to keep a quick eye open when Skem came into the library, check what he was taking out, and note it down.

When Marjorie first came, she expected that Skem would tell her what to do, but soon realized that Skem's way was to appoint the right person and let them get on with the job. She would say, 'This is what we're going to do,' and he would say, 'Oh, why is that, Marjorie?' If she had good reasons for decisions, he would be quite happy. This gave her a lot of confidence. He set up a library committee, chaired by Bob Gibson, of people who would be supportive to her, and provided a large budget. She remembers Skem saying, 'Spend away, Bob.' She enjoyed taking lists of books to H.K. Lewis, a technical bookshop in Bloomsbury, and cataloguing them when they arrived. Shelves were built in the spacious new room and she spent a happy summer with Skem and Gibson, aided by Katherine (aged 17, in her school holidays), sorting and shelving books from all over the department.

For a long time the (very dirty) historical books that had arrived from the Civils remained unsorted and uncatalogued because Marjorie's main responsibility was to organize the departmental library. The cataloguing work continued over many years, mainly during the summer holidays, during which time the collection was housed in special glass-fronted bookcases. Marjorie's *Catalogue of the Civil Engineering History Collection*, of which the books from the ICE formed the major part, was finally completed after her retirement and was published by the department in 1991.

A new era was beginning as an old one ended. On 25th October 1963 Karl Terzaghi died. Skem wrote the Obituary in The Times: 'By brilliant and highly original work on the mechanical properties of geological materials, Terzaghi created a new epoch in the science and practice of civil engineering.'

Ireland

Skem acted as external examiner at the Universities of Dublin, Cork, and Galway and made annual visits to Ireland between 1963 and 1965. He usually stayed with the resident professor, which was difficult in Dublin, as the professor's wife was, unfortunately, an alcoholic. More cheerfully, Skem and Nancy extended these trips to include short holidays at Kevin Nash's cottage on the wild and beautiful Connemara coast, at Renvyle, a village so small that it had only a postbox. All supplies had to be obtained at nearby Letterfrack. During this same period, Nancy made sketches for a wood engraving of Kinvarna Castle and Kinsale Harbour, Co. Cork, for that year's Christmas card.

Rankine lecture

The prestigious annual Rankine lecture which had been founded by the British

Geotechnical Society in 1961 was given on alternate years by an overseas and a British lecturer. In 1964, Skem was invited to give the fourth Rankine lecture, for which he chose the subject which had dominated his thinking for at least the preceding ten years: 'Long-term Stability of Slopes'. This was a really big event, for which Skem prepared over many months. He called upon the support of Bjerrum, inviting him to dinner to go through his lantern slides. He had to rework his draft almost completely a month before the date of the lecture in the light of what he was finding at the Walton's Wood slip on the M6. He also drew upon the work that John Hutchinson was doing at BRS on coastal landslips in southern England, which Skem describes as 'pivotal'. Hutchinson had found that, whereas current back-analyses of landslides were of initial failures, the great majority of coastal landslides involved renewals of movement on pre-existing slickensided slip surfaces, which had already experienced large shear displacements and that the effective shear strengths on these surfaces were much reduced in comparison with their initial values.

The commentary to the *Selected Papers on Soil Mrechanics* describes the significance of Skem's Rankine lecture. 'He demonstrated, for the first time, three highly significant results: first, that the strength on pre-existing shear surfaces in clay is distinctly lower than the fully softened value, due to re-orientation of the clay particles at large displacements; secondly, that this lower, or 'residual', strength as measured in the laboratory agrees well with back-analysis of reactivated landslides at Walton's Wood, in weathered Carboniferous shale, and at Sudbury Hill, in London clay; and thirdly, that, at Walton's Wood, it also agrees with the strength as measured on the shear surface itself.' The commentary goes on to explain, 'In the light of these discoveries, the Rankine lecture went on to re-examine the Jackfield landslide and the London clay cuttings in natural slopes, arriving at much improved interpretations.'

Bjerrum gave the vote of thanks, saying that Skem 'had guided every engineer through the jungle of apparently unrelated observations with which they were so frequently faced in soil mechanics. His ability to bring about order was based upon a number of unusual gifts. Imagination and intuition were needed to see possible correlations and to formulate alternative working hypotheses. Courage was required for rejecting irrelevant considerations while accepting those which were believed to be essential. The first and foremost requirement, however, was a thorough and intimate familiarity with the fundamentals of the problems to ensure that the conclusions were deeply rooted in sound basic principles... It is extremely fortunate that, in the present period of very rapid development, those interested in the subject have such a gifted person as Professor Skempton in their midst.' He also paid tribute to the extent of the influence that Skem had had on the development of the science all over the world.

In May Skem and Nancy visited Edinburgh for the International Congress on Large Dams, at which there was an extensive programme of visits for the 'ladies' to different parts of Scotland, glass works, and art galleries.

Sevenoaks bypass

In that same year, during construction of the Sevenoaks bypass in Kent, a drastic slip occurred. Even though the slopes were gentle, at three or four degrees, half the hillside suddenly started sliding. Alan Weeks, engineer for Kent County Council, called Skem urgently one Saturday morning. Skem rushed down to see the hillside still visibly moving. John Hutchinson remembers Skem taking a sample column of marvellous fossil soil for radiocarbon dating. Skem worked on this slip for two years with Nordie Morgenstern and turned it into a major investigation into the geological phenomenon known as solifluction, the downward slope movement caused by alternate freezing and thawing in periglacial conditions towards the end of a glacial period. The resulting paper, written with Alan Weeks, was published in the proceedings of a Royal Society discussion meeting on valley slopes which Skem organised in 1976. 'The Quaternary History of the Lower Greensand Escarpment and Weald Clay Vale near Sevenoaks', has become a classic among geologists and Skem regards it as one of his most important papers. This paper demonstrates Skem's competence in the sciences of geology, soil mechanics, and geomorphology. He reconstructs the various phases of periglacial solifluction and intervening periods of erosion from the Wolstonian to the present day and analyses the resulting geotechnical problems.

John Hutchinson considers that this paper brought to the attention of civil engineers, for the first time, the dangers inherent in disturbing low-angled and often innocent-looking slopes of periglaciated clays.

Sevenoaks was an example of the kind of site visit in England that Skem enjoyed more than anything. He was never happier than down in the bottom of a trench in his old donkey jacket, whistling cheerfully. Joyce Brown cites an incident during the Sevenoaks slip episode to illustrate how utterly undomesticated Skem was. One night Skem and Alan Weeks were down some trench until quite late. Alan invited Skem to his home for a bite to eat. He had just moved house and there was some chaos. His wife was perturbed that she had little to offer him except eggs. 'Oh, an egg would be lovely.' 'Would you like them scrambled or poached?' she asked. This seemed to puzzle Skem. After some thought, he said, 'Poached – that is to say, in a homogeneous mass.'

Montreal conference

1965 was the year of the sixth ISSMFE conference in Montreal. Skem was past president, and found this conference an emotional anticlimax after the excitement of the London and Paris conferences. Casagrande was president, and Skem found it tedious to be consulted on every detail by him. He remembers long discussions on matters that he feels Casagrande could easily have decided alone.

There had been considerable debate about the title of the society. After the Paris conference, Peck had written to Skem, 'the trend of the conference indicated that people have forgotten that foundation engineering is part of the title of the international organization'. (letter 11.8.61) Skem was advocating that rock mechanics should be included, while Bjerrum proposed the title, 'International Geotechnical Society'. By April 1963, a decision had been made and Bjerrum reported to Skem on his discussions with Casagrande and Terzaghi in Boston. Rock mechanics would be included but the name of the society would not change (letter 18.4.63). Skem remembers playing hookey during a particularly boring session at the Montreal conference to go and look at the surrounding countryside. The Mexican engineer, Zeevaert (whose work on Mexican clays he had discussed with Terzaghi in 1948), and his wife drove with Skem and Nancy to see a lake up in the hills. Skem also paid a visit to the Soil Mechanics Department at Laval University, Quebec, where his ex-Athlone student, Fred DeLory was then teaching.

Immediately after Montreal, Skem and Nancy moved onto Lisbon for the International Congress of Rock Mechanics, where Skem's subject was tectonic shear zones, and where they visited the newly completed Tagus River bridge, then one of the largest suspension bridges in Europe. They went on a long circular excursion right into the heart of Portugal, visiting the walled town of Batalha and Obidos with its beautiful ruined church.

In 1966, a symposium was held at the Institution of Civil Engineers on large bored piles organized by the Reinforced Concrete Association. At this meeting Skem presented a 'Summing Up' paper. This was based on work he had carried out in 1962 as consultant on earth slopes for the firm of Harris & Sutherland during the construction of the new University of Essex at Colchester. The site was in a hollow and there was trouble with the piles. It was thought they were building on London clay but the boreholes had not been made deep enough and it was found that there were layers of sandstone and clay.

It was around this time that Skem travelled to Sydney and Adelaide in connection with the construction of a dam in the Blue Mountains. He sent a card home to Katherine saying, 'Job going well. I have found a good bookshop in Sydney and made several purchases.' He came back waxing lyrical about the beautiful wrought iron balconies on the houses in Adelaide.

Cathedrals Advisory Committee

From 1964 to 67, Skem was a member of the Cathedrals Advisory Committee. He made a trip to Wells, to work on the stability of columns in the retrochoir. When the 14th-century builders had built the retrochoir, they had had to remove buttresses from the east end of the cathedral. When these buttresses were rebuilt, they came down on top of the columns, but not symmetrically, so in theory the columns should not be standing, but they had in fact stood firm for five centuries. Some consultant engineers had been called in and had proposed massive works. Skem did his investigations and strongly advised the Cathedral authorities to do absolutely nothing. (I guess this finding must have been greeted with enthusiasm, and not solely for engineering reasons.) Skem also visited Lincoln and Ely on behalf of the advisory committee.

Pisa

In 1964, the young Italian engineer, Carlo Viggiani, was attending a workshop at Gonville and Caius College, Cambridge, when he was thrilled to be taken by the then Director of Geotechnical Engineering in Naples, Professor Arrigo Croce, on a mission to Imperial College to recruit the famous Professor Skempton, author of 'The Pore-pressure Co-efficients A & B'. They needed him to join the Polvani Commission for the Tower of Pisa, which was charged with investigating the underlying geology and soil and looking into the cause of the lean of the tower. He enjoyed working with the Italian engineers. A snapshot taken in a beautiful cloister in Rome in 1965 shows him towering over the other members. Skem collected data for his Geological Society paper, 'The consolidation of clays by gravitational compaction'.

Viggiani sent me this vignette: 'When the subsoil investigations at Pisa began, there was a meeting of the Commission to examine the first samples recovered. We were on the green of the Piazza dei Miracoli with a long

The Polvani Commission with Polvani in the centre, Viggiani third from the left and Skem, '...towering over the other members.'

Kjellmann sample spread on a wooden table at the centre of a crowd of reporters, TV operators, and onlookers. Professor Skempton, his tall figure standing among the crowd, approached the sample, took a small lump of clay among the fingers and seriously fingered it, nodding gravely in assent. This produced some sensation among the presents [sic], and I was also very impressed. Many meetings later and, only at dinner, after a suitable quantity of wine had been consumed, I found the courage of asking Professor Skempton what had he done, what was the meaning of his gesture. And he answered, "None at all! But I have discovered that it is very effective in impressing people." I must admit that, since then, I have done myself the same performance a number of times, and always very successfully!'

The commission ended in the publication in 1970 of the Polvani proceedings in three 'Blue Books', which are still valuable documents on the tower. Subsequently, the Italian government set up another commission, this time to consider how to stabilize the tower. This Skem found to be a nightmare, dealing with complex and involved Italian politics, issues of national pride, and jealously guarded positions. This was just the kind of situation to drive Skem to distraction and he gave up the role in disgust. The IC's involvement in the very high-profile work on the Leaning Tower has been carried forward to great effect by John Burland, the current Professor of Soil Mechanics at IC, who, as well as being a consummate engineer, has the essential qualities of patience, charm, and diplomacy. In a recent lecture to the Friends of Imperial College, Burland demonstrated the quality of the soil underlying the tower very effectively by building up toy wooden blocks on top of a sponge. (They soon began to lean.) It is his work on Pisa which has made Burland internationally known and earned him his FRS. The lean of the tower is being gently and gradually reduced by means of removing material from the ground on the 'uphill' side. Burland receives faxes daily from Pisa, while the world holds its breath. (The complete and current acount can be found in 'Propping up Pisa', by John Burland in *Learning from Construction Failures: Applied Forensic Engineering*, edited by Peter Campbell and published by Whittles Publishing.)

In 1967, Skem was invited to deliver a special lecture at the European Soil Mechanics conference at Oslo. He chose as his subject the strength along structural discontinuities in stiff clays. The paper (with D.J. Petley, research assistant) was written several years after the Rankine lecture had drawn the geotechnical world's attention to the nature and problems of residual strength but, only a year or so after he had completed his work on Mangla, where the extensive sheared surfaces had posed such a serious threat to stability. 'The Oslo paper provided Skempton with the opportunity to develop further the ideas expressed in the Rankine lecture... He presented the first set of data on the strength characteristics of stiff clay joint surfaces.' (Commentary to *Selected Papers on Soil Mechanics*)

Chapter 7

A Leader of the Profession

Skem had just acquired a new female supporter at Imperial College. His right-hand woman and research assistant, Carlotta Hacker, had, in 1962, decided she wanted to travel round the world. She made plans with the wife of Peter Fookes but, when they met at Victoria Station to catch the boat train, they realized they had not agreed on which way round the globe they wanted to travel. Nevertheless, she set off and, as part of this mammoth journey, Carlo spent time at the Mangla Dam helping Binnie's.

Joyce Brown

Prior to her departure, Carlotta was instrumental in arranging Skem's next assistant, Joyce Brown. Carlo and Joyce had been friends since their undergraduate days at St Andrew's University, where they had both studied history and English. One evening Skem, Carlo, and Joyce had a meeting to discuss the job, at Carlo's room in Tregunter Road, just around the corner from The Boltons. The house was owned by a professional oboist, who had the ground floor, and the whole of the rest of the large house was let out as rooms for students and young professionals. This multi-occupancy would be unthinkable now in that extremely prosperous street. William Bennett, the flautist, was another tenant.

Joyce assumes that, following that Tregunter Road meeting, Skem went through the 'necessary channels', as, some weeks later, she was offered the job. During the 'interview', Skem enquired whether she knew anything about the type of mortar the Romans used in the construction of Hadrian's Wall. Knowing of her historical ability and northern connections, he sent her off to Newcastle to try and find a date for the first reinforced concrete cottage there. A task she undertook later for Skem was to discover the kind of cement used by Marc Brunel in his Thames Tunnel.

Like Carlo, Joyce very much values Skem's training in research methods. She says, 'He is a great checker and double-checker. He never writes his opinions or thoughts. This is due to a modesty in his character. He feels safe with facts, but not with opinion. He never writes anything he does not know to be accurate. If I trawled through 30 volumes and found nothing in the first 15, the temptation was to assume there was nothing to be found. Skem would keep me at it, and say, "No, it would be quite good if you could just look at the other half, then we could be absolutely certain." He asks the sharp questions which force you back on to clearer analysis of what you're doing.' Another research task was to go to the Patent Office to find any patents which showed iron in combination with concrete, to add to Skem's work on cements and early reinforced concrete. (At this time Skem was working on his 1962 paper, 'Portland Cements, 1843–87' on early cement technology, for the Newcomen Society.) She felt, as a non-engineer, that he gave her just enough of the technical information she needed to understand the task and no more.

Towards the end of Joyce's time with Skem, they wrote a joint paper, published in the Royal Society Notes and Records series, on John Troughton, the mathematical instrument-maker. Troughton had been apprenticed to the Grocer's Company, and was a member of the Smeatonian Society.

In 1972, Joyce joined the academic staff as lecturer in the Environmental Studies section. Joyce is a great friend of Marjorie Carter, the librarian Skem had poached from the Civils. They both essayed the poetical. Here is a piece by Joyce, 'On Clearing Professor Skempton's Desk', a nice geological analogy:

The foundations were well laid:
A layer of calculations made
In pre-history;
A sub-stratum of letters,
No mystery,
1958;
Grouts, composed of drawings,
Paper-clips and invitations;
The whole encastre
With reprints,
International, erudite, showing mastery,
Even hints of genius.
A topsoil of recent jottings
Once meaningful,
Now displaced.
Carefully I lift each layer,
Surprised at the penultimate
To find grey leather showing there.
It is the desk surface.

When you enter now, with purpose,
You will find him there,
Feeling, I must confess,
Exposed,
But benign still
And in repose
Above his blotting pad.

Skem always smoked a pipe and appeared oblivious of recent college no-smoking policies. His room was in a constant fug. Joyce found it hazardous, as he would knock out his pipe on the side of his metal wastepaper bin and occasionally set the contents on fire. He went on working, unconcerned, leaving Joyce to act as firefighter. His customary navy cashmere pullover from Harrods always had a hole or two in the front where burning tobacco had fallen on his chest. Joyce recalls how he often whistled Mozart cheerfully as he worked.

In October 1967, a postgraduate course in engineering seismology was announced, leading to the DIC. The star lecturer was Dr Ambraseys on earthquake engineering; Professor Bishop's contribution was on soil properties and earth dams, Dr Munro spoke on structural dynamics, together with Skem's input on stability of slopes.

Family history

Skem was by now in his mid-50s, a period of life when most people start to evaluate where they are and what they have achieved, and also to look back and see themselves in the context of their family story. During the late 1960s, prompted by the thought that elderly relatives may not be around for much longer to share their knowledge, Skem spent much of his leisure time researching, with his customary thoroughness, his own, and Nancy's family histories. This involved visiting surviving relatives in different parts of England and researching the towns, villages, and cottages where farmer and tradesman ancestors had lived, for example Buntingford in Hertfordshire and Twywell in Northamptonshire. I remember trips to damp churchyards, scraping moss off tombstones to read inscriptions, and the cajoling of vicars to allow access to parish records. Characteristically, he even researched the technologies various ancestors had used, for example, rope-making, and the reinforcing of concrete. He made an extensive catalogue of family silver and its origins, and of family photographs. This resulted in the two-volume family history that proved so helpful to me in writing the first chapter of this book.

The name Skempton is unusual enough for Skem to be interested in tracking down people with that name. For some time, he had known that there was a distant cousin, Alan Skempton, who worked at the BBC. In the early 1970s, Skem made a connection with his nephew, the composer, Howard Skempton. Howard's father, Ivor, was a physician in Cheshire and he attended annual meetings in London. One year the family decided to invite Skem and Nancy to meet them. Much later Skem attended concerts of Howard's music on the South Bank.

The anxieties about imminent mortality of the previous generation were well-founded because, in 1973, Skem's mother, Beatrice, became ill with what turned out to be pernicious anaemia. She briefly moved in with us at The Boltons but soon had to be admitted to the Victorian St Luke's Hospital (now demolished and replaced by the Royal Brompton). She died in 1974, to our great sorrow, and was buried in Weston Favell cemetery in a shared grave with her long-dead husband Alec. Skem has kept on his mother's house in Watford and still visits it every other weekend, to walk by the Grand Union Canal in Cassiobury Park, watch cricket on village greens, and have lunch in various favourite country pubs like the Royal Oak at Chipperfield.

Later, Skem had the distressing task of closing down his Aunt Olive's house in Paignton, where he had spent so many holidays. Her husband, Tom, had died in 1966 and, by 1982, Olive could no longer live independently and Skem had the emotionally stressful task of finding a nursing home for her.

At the other end of the generational scale, by the early 1970s, both Katherine and I were married and Skem had four grandchildren. He had approached the prospect of church weddings and walking with us down the aisle with trepidation and a marked lack of enthusiasm. But, in the event, he enjoyed the gatherings of family and friends at The Boltons in my case, and at the beautiful abbey at Malmesbury, Wiltshire, in Katherine's. Skem's morale was sustained by Silas Glossop, in his capacity as Katherine's godfather, plying him with a tumbler of whisky as he struggled into his Moss Bros morning suit at the Bell Hotel, Malmesbury. He regarded his grandchildren with benevolent but distant affection and left the grandparenting role in Nancy's capable hands. She was always a source of tactful advice and emotional support to Katherine and me in our mothering roles. I think we modelled ourselves on her example of loving firmness with young children. Katherine had studied geography at Bristol University and then local history

Judith's wedding… 'He had approached the prospect of church weddings and walking with us down the aisle with trepidation and a marked lack of enthusiasm'

in Leicester, and subsequently trained as a town planner. She and her teacher husband, Charles Sisum, set up home first in St Neot's, then in the village of Gringley-on-the-Hill, near Doncaster, where Skem and Nancy went every year for a summer holiday when Scarborough was no longer an option (Reg and Rosa having died in 1971 and 1977, respectively). When Katherine and Charles later moved to western Sheffield, on the edge of the Peak District, the summer holidays transferred there. Exotic holiday locations were not for Skem, except for conferences or consulting.

At home, he led a quiet and studious existence. Despite living in central London, he rarely went to the theatre or cinema. The occasional Early Music concert was the extent of his participation in London's nightlife. I remember going with him to lovely concerts held in the Raphael Cartoon Gallery of the Victoria and Albert Museum.

One annual social event that became a tradition was the Boxing Day party. The living room at The Boltons was packed with friends and colleagues and favoured research students, with, in the early days, Katherine and me pressed into service as handers-round of the vol au vents and Nancy's carefully prepared nibbles. Especially favoured friends, like Silas and Sheila Glossop, Eric and Vanessa de Mare, and Jas and Psyche Pirie were invited to stay on for a late lunch of spiced beef.

People I have spoken to have differing memories of dinner parties with Skem and Nancy at The Boltons. James and Anthea Sutherland, for example, remember one occasion when, at the end of the meal, Nancy led the ladies away from the table to the other end of the room, leaving the 'hard-drinking' men at the table. Although Anthea was very fond of Nancy, she was surprised that this Victorian custom was maintained, especially in a single room. The Sutherlands felt that Nancy very much subordinated herself to Skem, which she undoubtedly did. But they recognized and very much valued all her creativity. They were very appreciative of her making up a little kit for their daughter, who was, at one stage, very interested in bookbinding, for her project of binding a bible. After Katherine left home, Nancy, with time on her hands, did a teach-yourself course of cordon bleu cookery and menus at The Boltons became much more adventurous.

Mexico conference

The next ISSMFE conference was in 1969 in Mexico. Skem gave a State of the Art paper on stability of natural slopes with John Hutchinson, who, after his work on coastal landslides at BRS, had joined the department in 1965. This was the only collaborative paper where Skem did not do all the writing himself. His normal pattern was to write drafts of jointly carried out work, and send the draft to the co-author for comment. In the case of the 1969 papers with John Hutchinson, this was not possible, because John was a perfectionist and laboured over every detail. Skem found this somewhat trying. John remembers that, unlike now, a generous honorarium was given towards travel and hotel bills.

Laurits Bjerrum was the president of this conference. Ralph Peck describes the cacophony that was the closing banquet, 'the unexpectedly huge crowd, the deluge, the roving orchestras each creating more decibels than the other... Laurits had planned to introduce me as the new President. When the time came, he started his speech, but the roving orchestras were unaware of the ceremonies, and the attending guests could hear only the music. Nevertheless, Bjerrum was determined to continue his speech, and although I sat close by, I heard it not. Characteristically, he fought against the odds.' (Vignettes)

M4 slip

Skem's main work for Sir Alexander Gibb was on the construction of the M4 motorway near Swindon in 1970. Quite unexpectedly, very large slips occurred while cuttings were being excavated. It was unclear why this had happened. (There were similarities to the 1965 Sevenoaks bypass slip.) Skem followed his normal research method, investigating problems posed by real engineering problems. However it was for the Swindon job that he obtained his one and only research grant, to drill a borehole to establish the geological succession in the reactivated landslides. Skem found it was a shear-zone problem. He decided what should be achieved but the method of implementing the design for the remedial measures was devised by the resident engineer, Jim Watson, now head of Gibbs. The modifications involved huge cost. Skem never wrote up this work into a paper.

Honours started heaping up on Skem's head. In 1968, he was awarded a DSc (Hons) at Durham University and, at the Institution of Civil Engineers, the James Alfred Ewing Gold Medal. The recipient of this medal is chosen in consultation with the Royal Society and recommendations are made not only from the Civils but also from the Institutions of Mechanical Engineers and Electrical Engineers and the Royal Institution of Naval Architects. In 1972, he was awarded the Lyell Medal of the Geological Society of London in recognition of his contribution to the understanding of the behaviour of soils and of Quaternary geology. Skem regards his paper on 'The consolidation of clays by gravitational compaction' read to the Geological Society and published in their quarterly journal in 1970, as one of his most important papers. The Quaternary (that is, the most recent geological period) had previously been somewhat neglected by geologists, who, on the whole, preferred to study the older underlying rocks. The work in this area has subsequently been taken forward by John Burland and Dick Chandler, the current Professor of Civil Engineering at IC.

Burland is a great admirer of Skem but this does not prevent him having a realistic perception of Skem's less endearing qualities. Skem's impatience with people who bored him caused some difficulties. In 1974, after a conference in

Cambridge on the settlement of structures, Janet Roscoe invited Skem back to her house, together with Professor Jerry Leonard of Purdue University, USA. Jerry was questioning Skem in the American way. Skem got bored with this and cut him dead, Burland reports.

Burland also comments that Skem has never been a good committee man. In the mid-70s, he sat with Burland on the Council of the National Environment Research Council. He went week after week and rarely said anything – he was clearly out of his depth and eventually resigned in 1976. Burland suspects that it was this lack of interest in the politics and administration of academic life that prevented him gaining a knighthood at the time of his retirement.

Between 1974 and 1976, Skem was a vice-president of the Civils. Each year, one vice-president is groomed for presidency. This would never be Skem, because of his horror of committees and procedures. The vice-presidency was, rather, in recognition of his eminence in the field.

He did become a Founder Fellow of the Royal Academy of Engineering, which was set up to bring together the most distinguished British engineers drawn from all branches of the profession, to recognise the contribution of engineers to society, and to provide expertise on engineering-related matters. The inaugural meeting was hosted by Prince Philip at Buckingham Palace in June 1976.

Ambraseys considers that Skem is very good at making key appointments. He had an intuitive judgement and stored what he had heard in his 'computer brain'. 'He is a down-to-earth sort of person, very eloquent, not a theoretician,' says Ambraseys. When Skem brought in Alan Harris, designer of the Mulberry Harbours (the concrete docks that were towed across the Channel and used in the Normandy landings in 1944), as Chair of Concrete Technology in 1973, '...people were shaking their heads. His judgement was proved correct, however, as Harris set up an excellent section and is an eminent consulting engineer of the firm Harris & Sutherland, who, three years after this appointment, was awarded a knighthood for his services.' Another distinguished appointment of Skem's at IC was Professor (later Sir) Colin Buchanan, Chair of Transport and author of the celebrated report, 'Traffic in Towns'. Ambraseys says, 'Skem could see that transport was not theoretical, that you had to have a lot of field experience, that you needed to be able to talk to authorities and councils. Buchanan was a catalyst and set up a section which is thriving today.' (Personal communication)

The sudden death of Laurits Bjerrum was a great shock and source of grief to Skem. The occasion was the Rankine lecture of 1973 by Professor Bill Lambe of MIT. Bjerrum had travelled to London the day before and was staying at the Hilton Hotel but Skem had not seen him that evening. Professor Kevin Nash had gone to see Bjerrum during the morning of the lecture day and was shocked to find that he had died of a heart attack during the night. Skem was to introduce

Lambe. He heard of the death about an hour before the lecture. Lambe, who had come to the lecture hall in a bright suit and colourful bow tie, was all for cancelling the lecture but Skem argued that Bjerrum would not have wanted that. The great majority of the audience did not know what had happened, so Skem had first to announce the news of Bjerrum's death, in a skilfully worded introduction. Skem attributes Bjerrum's early death to his hectic lifestyle. As well as very actively directing his Norwegian Geotechnical Institute, he was in great demand internationally as a consultant. He jetted between Norway, America, and the Far East. He was a devoted family man, but often went from one job to another without going home. He had given the Rankine lecture in 1967, a year after Bishop.

Skem contributed a piece to a tribute to Bjerrum in *Géotechnique*: 'Laurits belonged to the extremely rare category of men who combine passionate enthusiasm with high intelligence and infinite capacity for hard work. His name will for ever hold an honoured place in the annals of soil mechanics and engineering geology... he will be remembered as a wholly admirable and delightful companion: gay, intelligent, generous, and warm-hearted, full of ideas (usually good ones) and possessed of quite extraordinary vitality. To meet Laurits and his wife was one of the great pleasures of my life.'

Peck summoned up for me a vivid picture of Bjerrum: 'He lived intensely, worked intensely, and took the greatest joy in his family and his friends. There were no halfway measures for him – he lived and worked with zest. He was a superb selector and organizer of talent, with the ability to generate support for and from his colleagues. NGI quickly became not only a great research organization but the world-recognized "finishing school" of soil mechanics. Yet, for all its technical achievements and its attraction for foreign scholars, its people were always Bjerrum's family. He was also a great showman with a keen sense of the dramatic. What to many would have been a dull technical subject became a mystery, a compelling detective story in his hands.' Bjerrum was as much an admirer of Terzaghi as was Skem. Their close personal relationship prompted Terzaghi to entrust all his papers to his enthusiastic young colleague and NGI now houses the Terzaghi library. (The Ralph Peck library has recently been added, in another beautiful room in the new NGI building in North Oslo.)

Skem never did any formal consulting work with Bjerrum but, on his many visits to London, Laurits would stay with Skem and Nancy at The Boltons and they would have long and intense discussions at IC about the fundamentals of soil mechanics. This was supplemented by an exchange of correspondence between them over the years on the themes of the properties of the clays of their respective countries, and sharing their ideas. When Bjerrum died, he was ten years younger than Skem and Skem's grief is intensified by the sense of what could still have been achieved. He was in his prime.

Mornos Dam

In 1976, the Mornos Dam was being constructed in Greece to increase the water supply for Athens via a very long aqueduct. There was a question of the stability of the hillsides immediately downstream of the dam. An increase of the water level in the dam would decrease the stability of the slopes and this instability would be accentuated by the risk of earthquakes. Skem worked on this dam with Nick Ambraseys, who describes a golden time, studying the geology, formulating the problem,and trying to discover the principles, sitting in the sun with sandwiches and a glass of wine, mulling over the morning's work. Skem enjoyed going to Greece and made several Greek friends. The contact led to a number of Greek postgraduate students coming to Imperial College.

The same year, Skem was called in by the consulting engineers to undertake a design investigation on the geologically interesting eastern bypass of the city of Bath. He travelled to Bath for several nights on several occasions, accompanied by Dick Chandler, who had been appointed lecturer at Imperial College in 1969. They worked with a senior man from the Geological Survey, G.A. Kellaway, whose geologically pronounced views did not always coincide with Skem's but whom Skem found 'likeable and enthusiastic'. This work led to a joint paper with Chandler and Kellaway for the Royal Society, 'Valley Slope Sections in Jurassic Strata near Bath'.

University of California

The same year, 1976, out of the blue, Skem was given the honour of appointment as Foreign Associate of the National Academy of Engineering of the USA. This is an honorary appointment, in acknowledgement of achievement in the field. Skem imagines Ralph Peck was influential in nominating him.

Bn contrast the post of Hitchcock Foundation Professorship of the University of California was a real job. The professorship is an annual appointment at the university and can be in any subject. Skem went to Berkeley for a month in 1978 and gave a short course of lectures in the Civil Engineering Department. He also gave two lectures in Los Angeles to the university as a whole, which were huge events, arranged by Harry Seed, Professor of Soil Mechanics at Berkeley. One was on the Sevenoaks geology and the other on early cast-iron bridges. He had given this lecture once before while en route to California. In SoHo in New York, there were a large number of extraordinary iron-fronted buildings. It used to be a commercial area and they were mainly warehouses. An energetic lady, Margot Gayle, set up a conservation society, 'The Friends of Cast Iron', to prevent their demolition, and it was she who invited Skem to give the lecture. Margot had an apartment in Greenwich Village and she knew all the good places for coffee and lunch. Skem had a

DONNELL LIBRARY
20 W.53rd Street, NYC
WEDNESDAY, NOV.IST
6:00 P.M.
FREE

FRIENDS
OF
CAST-IRON
ARCHITECTURE
Present An
illustrated lecture by

Prof.A.W. Skempton

Renowned British scholar
discussing early structural
uses of iron

**CAST IRON
BRIDGES 1777-1839**

Co-sponsored by
The Victorian Society
Society for Industrial Archeology
Society of Architectural Historians

Leaflet announcing Skem's lecture on cast iron bridges.

thoroughly enjoyable two days with her and her right-hand man. The lecture was held in the Donnell library, W 53rd St.

Skem was most impressed with the beautiful campus, high-quality architecture, and academic excellence at Berkeley. He occupied the room of Richard Goodman, biographer of Terzaghi, who was on sabbatical leave at the time. He found the month very tiring. His bedroom in the university guesthouse was on the corner of a building round which traffic swirled day and night. He arranged to move to a smaller internal room to get some peace. He was almost continuously on show, partaking of the Americans' famous and inexhaustible hospitality. A dinner here, a trip there, drinks somewhere else every day, and he came home to England exhausted.

He found refuge from all this sociability at the Sutro library in San Francisco. Sutro was a late-19th-century mining magnate and book collector, who had bequeathed his huge library to the City of San Francisco. Among other riches, he had obtained a collection of early printed reports collected by Sir Joseph Banks, a Lincolnshire landowner, who had been a famous President of the Royal Society in the late 18th century and who had purchased reports by Smeaton and Grundy and others on fen drainage.

It was at this time that Skem's historical research centred on the work and engineers of the 18th and early 19th centuries. He was compiling his bibliography, *British Civil Engineering Literature 1640–1840*, so the opportunity for research in the Sutro library was very valuable. The book is an important addition to the bibliography of the subject, and is now known in the book trade simply as 'Skempton'.

Julia Elton

On his return to England, in order to extend his growing collection of antiquarian engineering books, he often visited the Great Russell Street shop of Ben Weinreb, the only dealer in London to specialize in this subject. According to him, this is where he met Julia Elton, who was young enough to be his daughter but who became one of his closest friends. Julia's father, Sir Arthur Elton, of Clevedon, in North Somerset, had assembled a formidable collection of books, works of art, and commemoratives relating to the Industrial Revolution and had edited the book *Art and the Industrial Revolution*, of which Skem has a signed copy.

On his death, which was devastating for Julia, Ben Weinreb was asked to handle the passing-over of the collection to the nation in lieu of death duties. Julia prepared a catalogue of the pictures and artefacts to assist Ben in this task and, as a result, he gave her a job. Ben's main specialism was architecture but he had also amassed a stock of books on engineering and Julia's background gave her a natural interest in this field. Ben insisted that she join the Newcomen Society and she began gradually to educate herself in engineering history. I think she also joined the Newcomen Society as a way of entering her father's world. She is now well established as a specialist bookseller in her own right but this was far from the case in 1976, when she met Skem.

However, Julia's version of how she and Skem met is different. She says that she had gone with her mother to one of the famous Sunday afternoon salons at the house of their friend, Jean Gimpel, in Drayton Gardens. Gimpel is 'an extraordinary Frenchman who was fascinated by mediaeval technology,' and is known for his books, *The Cathedral Builders*, and *The Mediaeval Machine*. Skem denies that he was even there, saying it is the kind of gathering that he would dislike but the meeting made an unforgettable impression on Julia. 'There was this heroic man, and I heard my mother saying, "My daughter is very interested in the history of engineering." I was living alone in South

Kensington, very close to the Skemptons, and Skem said, "Perhaps you would like to come and have a cup of coffee with my wife and me one evening." She politely accepted, knowing she was far too shy to follow up the invitation, but, a few days later, Skem issued a formal invitation, almost an order, to come to dinner. 'To this day I have never understood why Skem took such a kindly interest in me. I was very young, very ignorant, but I had landed this miraculous job at Ben's and Skem, I suppose, took a fancy! When I turned up to dinner, I was just sufficiently on the ball to have grasped the name of an 18th-century engineer/architect called Robert Mylne and I took along a group of little books

Skem at his desk. A 'formidable figure' felt Julia Elton.

on fen drainage which had his ownership inscription in them. Skem, bless his heart, assumed I knew everything about Robert Mylne. The whole evening was something of an ordeal. Skem and Nancy were terribly distinguished and had a great presence, both together and apart. Neither of them had any small talk and I simply did not know what to say to them. After dinner, we had coffee and I did my best to involve them both in conversation – Nancy with her bookbinding – but then Nancy went out of the room and Skem settled down and started to talk about Mylne. I hadn't taken on board that, actually, of course, Nancy was perfectly happy if Skem was happy. Anyway, after that, he rather took me under his wing. When he was on his way to or from the British Library, he loved to come to the office and see Ben. Having this hero arrive on our doorstep, produced by me, was a plus with Ben! He would sit in the Best Bookroom with his pipe while I got out books to show him and Ben scurried round offering sherry. One icy winter day, he fell flat on his face in Great Russell Street and came into the office as white as a sheet, trembling in every limb. Ben lent me his big white Jaguar so that I could drive him home. He used to take me out to lunch in a local Greek restaurant, where he would tell me about his latest piece of research, either in engineering history (he was then writing the biography of William Jessop) or in soil mechanics. He has a gift for making even the knottiest foundation problem sound perfectly simple and he would draw diagrams for me on a napkin until I understood what he was talking about. I still wake up in the night and wonder how I could have been so lucky. If I say to anyone in the engineering world, "Well, actually, Professor Skempton is a great friend of mine," they all fall flat in amazement!' In fact, Julia and Skem do have the same sort of historical imagination and like the same sort of books and she, like Sonia Rolt, can make him laugh.

One evening, when he was at dinner with Julia and her partner, the structural engineer Frank Newby, they discovered that Frank, also a passionate book collector, had a copy of Smeaton's 1766 report on the Lee Valley navigation, in a composite volume of material. Until that moment, Skem had assumed that the report existed only in manuscript and was wildly excited to find that there was a printed version. ("The report of John Smeaton, Engineer, upon the New-making and Completing the Navigation of the River Lee, from the River Thames, through Stanstead and Ware, to the Town of Hertford." London 1766. Skempton biblio. no.1306) Frank gave it to him and, in return, Nancy rebound the contents of the rest of the volume.

Miraculously, a second copy of this rare item subsequently came up for auction at Sotheby's. Julia went to bid for it but was so astonished when it went for such a large sum that she felt she couldn't buy it. She was so overexcited that, later in the sale, she got in a muddle and bid on the wrong lot, buying the Countess of Blessington's memoirs instead of some esoteric volume on Highgate which Ben Weinreb had wanted her to get for his London collection. Skem thought this was a great joke. Julia says, 'Quite often, when we were sitting in some restaurant after a Newcomen Society meeting he would say, "Let's have the story, Julia," and we would tell it as a kind of duet. When I got to the part about bidding for the wrong lot, he would burst into gales of laughter.'

Skem tells the story that soon after he and Julia had made friends, Lady Elton, somewhat imperiously, invited herself round to The Boltons to give him the once-over, to check whether he was a suitable friend for her daughter.

Tokyo conference

Skem did not go to the eighth ISSMFE conference, held at Moscow in 1973. They had become such huge events that he perhaps felt they lacked the intimacy and excitement of the early international conferences. But, in 1977, came the ninth conference in Tokyo, where Skem gave a special lecture, which summarized the development of understanding of first-time slides in cuttings in brown London-clay slopes. Nancy did not accompany him to Tokyo. He recalls meeting the leading Japanese engineers and attending what was, for him, a very trying experience, a formal dinner where he had to sit cross-legged almost at floor level at a low table being ceremoniously served by geisha girls in traditional dress and full white make-up. At home, Skem is notorious for finishing his food before the rest of us and, as we enjoy exchanging the news of the day, he impatiently drums his fingers on the table (a reminder of his jazz drumming days), keen to return to whatever he is working on at the time. On this evening in Japan, he had to endure a succession of tiny courses, each more inedible than the last, interspersed with interludes during which the geishas did their stylized traditional dancing. The meal lasted for many hours and, while I am sure he appreciated the hospitality being offered, it must have been a test of physical and mental endurance for him.

Newcomen Society

The Newcomen Society for the Study of the History of Engineering and Technology had been founded as far back as 1920, with its base at the Science Museum in South Kensington. It produces a learned journal, which had published Skem's first historical paper, on Alexandre Collin, in 1946. Then came his papers on river navigations, William Strutt, the Sheerness Boat Store, the steam dredger, and many others. He received the Newcomen's Dickinson Medal in 1974 for his contributions to the history of engineering and gave the Memorial Lecture on "William Chapman, 1749–1832." He was president of the society from 1977–79, an honour he valued highly. The role of president involves chairing the monthly meeting where papers are given. These papers can be on all sorts of engineering-related subjects. Sonia Rolt has the impression that Skem never missed a council meeting, even though they were, in Sonia's words, 'sleep-inducing'. 'There were delightful people, the old brigade, who used to come and sit in the front row at lectures and, within five minutes, no matter who was giving the lecture or what the subject was, they were asleep.' Dr Stanley Hamilton, 'a splendid figure, a lovely chap', who had influenced Skem towards engineering history in the early days at BRS, was one of these. Council meetings are very different nowadays, says Sonia, 'full of vibrant personalities like Julia Elton'.

Highlights of the Newcomen year are the annual summer visits, which were instituted very early in the society's history. Sonia Rolt, who became a member after Tom's death, thinks that some particularly enjoyable ones were organized during Skem's presidency, although he claims no credit for this. They were arranged by 'the local fellow' in whichever area they were visiting, he says. Sonia reported, 'The period when Skem was president was altogether enjoyment, because he was very laid back about it all. He didn't do a great deal of the arranging. Some people got knotted brows about it, but not him.' The Newcomen trips lasted four to five days and up to 50 people came on them – sometimes two coaches were needed to transport them around. They usually stayed in university halls of residence and a formal dinner was held.

Sonia tells some hilarious tales of these summer visits. Once, at Wrexham, the solemn dinner was suddenly enlivened by 'a group of little Welshmen in pink suits with bow ties launching into a barbershop quartet. An extraordinary party. Then there floated in some lovely girls in diaphanous draperies. I remember Skem's total amazement when some diaphanous lovely came and hovered about, practically perching herself on his knee. One had a hysterical evening. There were these crusty distinguished gentlemen, then suddenly, on the Isle of Man, people playing fiddles and dancing about. They were forced to dance as well. Like elephants at play.'

Julia tells how Skem's ready sense of humour enlivened so many of these visits, but she also gives an example of his lack of patience with what he saw as small mindedness. At one Newcomen visit, this time to Sunderland Harbour, he was furious that the group's attention seemed to be focused on some steam crane or other, ignoring the great harbour itself, the remarkable achievement of the Shout family of engineers. He apparently summoned a taxi and dragged Nancy to the station to catch the next train back to London.

Sonia was sorry that Skem did not come to visit Tom when he was dying, in the spring of 1974, but acknowledges that 'one of the ghastlinesses of prolonged cancer is that it reduces you both physically and as a person. It's the one image that we try to shut out of our minds, the Belsen image. Skem would not have wanted to deal with that. But he couldn't have been kinder to me when I was on my own. He did it very well. I was in fearsome grief. He took it as read, treated me gently, and did not discuss it. Some people felt they ought not to mention Tom, but Skem did manage to speak easily about him, not making an issue, it just came naturally to him.'

Julia Elton takes up the story: 'So in the year Skem became president of the Newcomen Society, Sonia and I were these two vulnerable little creatures. He was so kind to us. For the two years that he was president there would be these dinners after the papers. We would get to the restaurant and he would say, "Sonia, dear, come and sit by me. Julia, dear, come and sit here." Skem would have Sonia and me sitting there with the speaker and we figured out that probably he needed some help, partly because Nancy by no means always came. I think she came when he was lecturing but she certainly didn't come to the rest. Skem would run out of questions, half the papers were about steamy subjects – nothing could be less interesting for him. So Sonia and I did an awful lot of wagging our tails. I remember saying in some crass way, I can't remember the context, perhaps at a dinner at The Boltons, "Sonia and I are Skem's camp followers." Nancy pulled herself up to her full height and said, "Skem does *not* have camp followers", and me feeling, Oh, they will never speak to me again. Any way we somehow got over that one.'

Another memorable Newcomen summer visit was to the 'Engineerium' in a huge old pumping station just outside Brighton, where there are beam engines and a fabulous collection of models, for example, a model of the Rocket locomotive made for Robert Stephenson. The Engineerium was started by Jonathan Minns, who used to own a model shop in Hollywood Road, just around the corner from The Boltons. Skem particularly enjoyed the Brighton trip because Geoffrey Binnie was there with him. They had become great friends after Mangla. They went on trips to Ramsgate Harbour, to the open-air Weald and Downland Museum, and to look at 'hammer-mill' dams for ironworks that used to straddle the border between Kent and Sussex. Binnie wrote a book on *Early Dams in Britain*, which includes some illustrations of these dams. Skem helped him with this book and Binnie acknowledges this contribution.

Sonia remembers another nice time at 'that sort of spiritual home of uplift in Newcastle', which Skem explains is the Newcastle Literary and Philosophical Society (Lit. & Phil.), one of those 19th-century local philanthropic societies started by active and wealthy industrialists, specially in Northern England (there is a famous one in Manchester where the great chemist, John Dalton was a leading member). At Newcastle, there is a magnificent library, which Skem compares to the Royal Institution library, and a lecture theatre, where the Newcomen Society held an evening lecture on one of the summer trips.

PCL history lectures

Apart from the Newcomen, even in the mid 1970s, very little was known about the subject of the history of engineering; it was not a popular topic for study. But, during the late 1970s, the Polytechnic of Central London (PCL, previously Regent's College) held a ground-breaking series of history lectures, which were important in raising awareness and interest. These were grouped into three series, in 1976, '77 and '80, and were organized by Roger Bishop and Frank Weare of PCL.

Skem was one of the lecturers in each of these series of seven lectures. In

1976, he spoke on the introduction of cast iron, with Rowland Mainstone talking on structural form and James Sutherland on wrought iron. In November 1977, there was a second series, this time of seven biographical studies. Frank Weare did Telford, Trevor Turner did Smeaton, and Skem spoke on Jessop, whose life he was busily researching for his book. John James did Sir Charles Fox (who built the ironwork on the Crystal Palace), and Joyce Brown spoke on Proby Cautley of Ganges Canal fame. The talks were extremely well attended, with people squashing into the small Room 20 at the Poly. Julia Elton went to the second series 'under Skem's wing'. (There was a third series of lectures in 1980, when Frank Newby spoke on structural design, Alan Harris on concrete, and Skem discussed 'classical soil mechanics'). One keen audience member was Malcolm Tucker, whose comments speak for many who were present. 'The 1976 series was an eye-opener for me, since I had previously taken a strong interest in the archaeology of structures, but had not got to grips with the formal history, particularly of structures long vanished which complete the picture.'

Not only were the lectures marvellous, but they acted as a kind of catalyst bringing together a group of friends who now form the solid core of the Institution of Structural Engineers' History Study Group. (See Chapter 8)

AIA and PHEW

Skem is involved in two other organizations that involve themselves in the history of engineering. The first of these is the Association for Industrial Archaeology (AIA), which had been formally established in 1973 at a meeting in the lecture hall of the Institution of Civil Engineers, when Tom Rolt was elected as the first president. A subsequent meeting of AIA took place at Rolt's cottage at Stanley Pontlarge, Gloucestershire, because he was too ill to travel to London. Everyone present signed the Rolts' visitors' book. After Tom died the following year, Professor Angus Buchanan took over the office. One of his first tasks was to set up a memorial lecture programme for Tom Rolt. The idea was that a distinguished person in the history of technology should be invited every year to give a 'keynote' lecture to the annual conference of the Association, and this would be known as the Rolt Memorial Lecture. Buchanan invited Skem to give the first lecture in Durham on September 14th 1975. He spoke on 'The Engineers of Sunderland Harbour' (later published in the *Industrial Archaeology Review*. Vol. 1, No. 2, spring 1977). Skem and Nancy were looked after by Buchanan and his wife in Durham for some of their time at the conference.

Buchanan was a fellow member, with Skem, of the council of the Newcomen Society, and he is now Director of the Centre for the History of Technology at the University of Bath. He pays tribute to Skem: 'I have always known him as a person of great authority, good humour, and decisiveness, as well as being an excellent scholar.' Sonia and Julia accompanied Skem to subsequent AIA meetings. Sonia says that, if they were on the 'top table' with him, there could be quite a lot of bad feeling about it among other members of the society, because 'they also would like to have the light of Skem's countenance shining upon them'.

The second historical organization was the Institution of Civil Engineers' 'Panel for Historical Engineering Works' (PHEW). This was set up in 1971, and has 12 panel members, one for each region. Professor Roland Paxton from Edinburgh is the present chairman, then there is Roger Crag in the West Midlands, Gardiner in the East Midlands, and Bob Rennison up in Newcastle. Each regional member organizes summer trips to interesting places, which are so similar to the Newcomen visits that Skem, looking back, has difficulty remembering which is which. He was chairman of the panel for eight years, between 1981 and 1990. Mike Chrimes remembers him being very firm if he was not happy with the way a meeting was going.

The panel produces the series of little books, *Civil Engineering Heritage*, on each region of the country, the engineering equivalent of Pevsner's famous *Buildings of England*. Roland Paxton is working on the Scotland edition and Denis Smith on the volume for London and the Thames Valley. James Sutherland is critical of the format in which these books are produced, which he thinks militates against them becoming as popular as they deserve. They won't really go into your pocket but they do not have the solidity of the Pevsner volumes. On the other hand, they have good illustrations, maps, and useful indices. Were they to have wider circulation, they would contribute to a raising of public awareness about the importance of engineering.

Landmarks in Early Soil Mechanics

At the seventh European conference on soil mechanics at Brighton in 1979, Skem gave his famous paper, 'Landmarks in Early Soil Mechanics'. This dealt with the development of the subject in the 150 years between Coulomb and Terzaghi. His introduction sets out his understanding that 'the subject existed as a set of somewhat isolated topics, such as earth-pressure theory and practical knowledge of slips in clay slopes, with little correlation between field observations and theoretical analysis and lacking, above all, the unifying principle of effective stress.' He goes on to deal with some of the more interesting theoretical and practical contributions of that period, such as Grundy on earth dams, Poncelet on the design of retaining walls, Jessop and Telford on the Caledonian Canal, and Rankine's *Manual of Civil Engineering*. He also refers to the work of Osborne Reynolds in the 1880s on the friction between earth particles.

This was the cause of further upset with Rowe at Manchester University. Rowe had written a paper on Reynolds and felt that Skem had completely ignored his work and was wrong in saying that it was the work of Skinner (of IC) that had moved thinking on from Reynolds. Salt was rubbed into the wound by a passage in the published version of the paper, in which Skem wrote, 'In cohesionless materials, phi is controlled almost exclusively by particle shape and packing, and hardly at all by the co-efficient of friction between the grains, as shown in a beautiful series of tests by Skinner (1969).' Rowe wrote Skem a furious letter (one of those which Skem declined to deal with), accusing him of undermining his life's work. Despite John Burland's best efforts and extensive correspondence from Skinner, Rowe vowed that he would never again have anything to do with IC.

The Brighton conference took place in a riding school built for the Prince Regent, which has been transformed into the Dome complex behind the Pavilion. Nancy and he stayed in a very comfortable hotel next to the Grand on the Brighton seafront. It was this conference that led to John Burland coming to Imperial College.

The following year, Bishop, who was Professor of Soil Mechanics under Skem, finally retired through ill-health. Peter Vaughan, who had, in fact, been running the section throughout Bishop's incapacity, generously encouraged Burland to apply for the professorship, even though he himself was also applying. The fact that Burland got the job rather than him was difficult only for a time, such is the easy-going nature of Peter Vaughan.

Jessop book

In the early 70s, the question had arisen of Tom Rolt writing a book on Smeaton's pupil, William Jessop. Jessop's papers have not survived, so it would be a major task, and by then Rolt's health was giving out. He passed his notes on to his great friend, Charles Hadfield, author of an important series of books on the canals of the British Isles. He had always admired Jessop, and asked Skem to help with the book. They had great struggles to get material. Skem searched county record offices in the places where Jessop had done jobs. For example, he and Nancy spent a week in Bristol looking at records of the floating harbour. Nancy made notes, copied drawings, and generally acted as research assistant. He stayed with Katherine at Gringley-on-the-Hill while researching the records of the clerk of drainage board at Bawtry, for information about Jessop's work on fen drainage. He looked at accounts in the East Riding record office in Beverley about the Holderness drainage scheme. He spent two to three weeks at the Port of London Authority researching Jessop's work on the West India Docks.

After all these labours, their preface to the book, somewhat plaintively, suggests, 'It is, we think, a useful rule that, if a man desires his work to be well regarded after his death, he should take care to leave in reliable hands a large collection of personal papers into which historians and seekers after doctorates can happily burrow.' Hadfield wrote the canal chapters and Skem the others. Controversy arose between them because of Charles' passionate advocacy of the importance of Jessop's contribution in relation to Telford's in the building

Charles Hadfield, from Charles Hadfield.
Canal Man and More *by Joseph Boughey*
(Sutton Publishing).

of the Pont Cysyllte aqueduct on the Ellesmere Canal. Charles' biographer, Joseph Boughey, reports that, 'He did amend the chapter dealing with Telford and Jessop, at Skempton's request, but continued to express suspicions about Telford's autobiography.' (Hadfield still thought Jessop's contribution was overlooked in favour of the more famous Telford.)

Julia Elton tells a tale of a Newcomen meeting which took place around the time when Charles and Skem were just finishing the Jessop book. 'There was a terrible time when Skem and Charles Hadfield behaved mortifyingly badly. There was this young man who was giving a lecture on Pont Cysyllte. Charles and Skem had become entwined with Jessop to a quite ridiculous degree, Charles Hadfield actually more so than Skem but, at the time, they were in love with Jessop, I can't think of any other way of putting it. This young man, Philip Cohen, was giving a paper ("Origins of the Pont Cysyllte Aqueduct", Trans. Newcomen Soc. Vol. 51) and the burden of the paper was the fact that there was an earlier iron aqueduct in Merthyr Tydfil. He was a plain and unappealing young man but Skem could just as easily have been as nice to him as he was to me. Anyway, he gave this paper and Skem and Charles Hadfield went for the jugular. It was terrifying. Wham, wham. First of all, Charles Hadfield gets up and lays into this helpless, hapless young man and sits down, and then Skem gets up with his hands, like he does, in his coat pocket and he just tore the young man to shreds. At the dinner, I said something soothing in a misguided attempt to pour oil on the extremely troubled waters.

The bottom line was that the young man was right.' I asked if Skem had ever admitted this. Julia said that 'it was patched up in some mysterious way. They were very graceful, they wrote to him (acknowledging an error in the Jessop book) and then it was printed in *Transactions*. Something very graceful came out of it. Skem *is* very graceful.'

Despite this upset, Skem very much enjoyed the process of writing the book, which took two years. He liked Charles and admired him as a very hard worker. Charles lived from his writing and jointly founded the publishing firm, David and Charles, at Newton Abbott. Skem had known him for some time through the Rolts. Charles had lived at South Cerney, near Stanley Pontlarge, but, in 1977, he and his wife, Alice Mary, were living in Little Venice, in London. Apart from Skem's visits to Little Venice to discuss the Jessop book, the Skemptons and the Hadfields often visited each other for friendly suppers. When Alice Mary died, Charles was disturbed at what he found in some of her papers relating to her life before she met him, other relationships she had had. He never recovered from her death and went voluntarily into a nursing home, saying he wanted to join Alice Mary. Skem found this very upsetting.

The Jessop book is full of very thorough research and returns to original sources but a criticism of the book could be that, although it is very detailed, giving, for example, the prices Jessop charged and quotations from his correspondence, it never really brings the engineer to life. Geoffrey Binnie, however, reviewed the book favourably for the *Newcomen Bulletin*: 'The story that the authors have unravelled is the most exciting one during a great period in the development of British and Irish waterways and docks that has come to light for many years and it will enable Jessop to take his rightful place.'

Telford at Ironbridge

Julia Elton tells the tale of the Telford conference at Ironbridge Gorge in April 1979. Skem was giving an important lecture on Telford's 600 ft span cast-iron design for the new London Bridge. Ben Weinreb had copies of Telford's original competition drawings. They were huge plates, folding out five times. Skem asked Weinreb whether he could take these drawings up to Ironbridge to supplement his slides. Julia travelled up to Shropshire with Skem and Nancy. 'It was a ghastly new car, an Austin Princess. Skem couldn't make it work. Every time anything creaked, he said, "There's a noise, oh dear, oh dear." Nancy was sitting, totally impassive, and I was in the back with the drawings and suitcases. We stopped for petrol and there was Skem faffing about saying, "Oh dear." (It was the first time he had encountered a self-service pump.) I finally said to Nancy, in a polite kind of way, "Should I get out and help?" She said firmly, "No, Skem must learn to do it for himself." We finally got to Ironbridge, and gosh it was fun.'

Sonia Rolt, who has always been very involved with the Landmark Trust, which purchases and renovates endangered buildings of architectural interest to let as holiday properties, had taken the Landmark Trust house in Ironbridge. Neil Cossons was then Director of the Ironbridge Gorge Trust. Julia goes on 'Sonia had sent a shopping list, and Neil's heavenly wife had been to Tesco and brought all the food. She arrived on the doorstep, laden with enormous bags, and the four of us moved in [Skem and Nancy, Sonia and Julia]. We partied and partied. I remember standing in the kitchen with your mother, and saying rather shyly, "Could I call you Nancy?" By this time, I had known them both for some years.' While at Ironbridge, they visited the famous slip at Jackfield which Skem had written about in his Rankine lecture. (Sir Neil Cossons, recently knighted, later became Director of the Science Museum, and is now the head of English Heritage.)

Mam Tor

In February 1977, after heavy winter rains, there was a very large movement on the ever-unstable road leading up from Hope Valley over into Edale in the Derbyshire Peak District at Mam Tor. The road had been built in 1810 and had been rebuilt on several occasions because of slides. In 1979–80, Derbyshire County Council was considering whether or not the slip could be stabilized and called on Skem to advise. He worked on the job with Dick Chandler. They made several visits together over a period of a year, staying at Matlock, and driving up to the site daily with Derbyshire's engineer, Leadbeater, and making numerous boreholes.

It was a very ancient slip, with the road going right across it. The slide was deep-seated, especially in the upper part. Skem and Chandler tested this soil. They were much aided by the fact that detailed records of both rainfall and landslide movements had been kept since 1915. They were thus able to relate the landslide movements to rainfall during that time, movements having occurred on 20 occasions, all in wet winters.

In advancing, the toe of the slip had moved over the original soil without disturbing it. This undisturbed fossil soil contained organic matter which, with the help of Dr R. Williams of Birmingham University, who carried out radiocarbon dating, established that the soil in the toe was 3000 years old. The contact with Williams was brought about through a friend of Skem's, the geologist, Professor Fred Shotton. From this dating, the historical records, and the changing geometry of the slip, they concluded that the slip had started around 4000 years ago. Skem had made use of radiocarbon dating of clay at the Sevenoaks slip back in 1968 and these two instances were pioneering the use of carbon dating of this type in soil mechanics. Interestingly, there is another slide in neighbouring Edale in the same geology that had started 6000 years ago and that was completely stable. There had been no trouble at all with the road in that valley.

Derbyshire County Council was never keen on the idea of stabilizing the Mam Tor road. They realized it would be enormously costly. The road has been abandoned at the slip and traffic now has to revert to the original circuitous route over Winatts Pass. Since 1978, the slip has moved a further six to eight feet, fully justifying the council's decision.

There was a long delay between Skem and Chandler doing the work and the publication of the resultant paper by the Royal Society in 1989. This was because the Carsington dam failure intervened. As at the Sevenoaks bypass, Skem found the geology surrounding the Mam Tor slip absolutely fascinating and the paper only mentions the actual road, as it were, in passing. The paper represents a development of his thinking and puts Mam Tor in the context of other large slides in the Pennine area. Skem submitted it to the Royal Society because they publish papers in an attractive large format with good photographic reproduction and have a tradition of publishing papers on Quaternary geology.

The two cultures

Meanwhile, with my children now at school, I decided to train as a social worker. This career choice caused Nancy and Skem some consternation. Skem had always hoped I would choose librarianship but, failing that, they thought that, with a scientist as a father and an artist as a mother, they had, in true C.P. Snow 'two cultures' style, knowledge of all areas of human endeavour. They understood nothing of social work and found it difficult to understand why I wanted to work with the poor and underprivileged in society. A few years later, when Katherine abandoned her town planning and trained as a speech and language therapist, they had prior experience of a daughter doing something different but, as speech therapy is a 'profession allied to medicine' they found it not quite so outlandish.

Every summer, Skem and Nancy spent their summer holidays staying with Katherine and her husband Charles. When they were at St Neots, Skem spent time studying the fen drainage in that area. They then moved to Gringley-on-the-Hill in Nottinghamshire. Whenever Skem stayed with them there, he studied the Chesterfield Canal and Jessop's 1816 scheme for draining the 'carrs'. Their next move was to Sheffield, where Charles took up the headship of a large comprehensive school. For Skem, this was gratifyingly near the Mam Tor slip, as well as walks in the Derbyshire Peak District.

Nancy's occupations

Skem and, perhaps even more, Nancy, became active members of The Boltons Association, the residents' group for the area. In the 70s The Boltons was still home to academics and a mixed, if prosperous, group of people. (Now only the super-rich can afford houses there.) They made friends with one of these, Professor Darrell Forde, an anthropologist, whose wife, a doctor, was related to

Charles Singer. When Skem and Nancy moved to The Boltons, Singer had mentioned that Professor Forde lived opposite and the couples subsequently became firm friends. Other distinguished academics living around the garden of whom Skem was aware were Professor Andrade, Chair of the Royal Society Library and Director of the Royal Institution and his older predecessor at the RI, Sir William Bragg.

The Boltons had been designated a conservation area by the Borough of Kensington and Chelsea in January 1970. Soon after this, another neighbour, Mrs Austin-Smith, set up The Boltons Association. Nancy expressed some of her rural longings by becoming deeply involved in a tree survey on behalf of the association, building up information on the number and variety of trees in local streets and garden squares and working with the local council to list the best specimens to prevent their destruction by development. This work is acknowledged in a document relating to The Boltons Conservation Area. Nancy used to love to visit Kew Gardens, often with her friends, who were wives of Skem's colleagues, for example, Mel Nash (wife of Professor Kevin Nash), and the wife of John Francis, Professor of Hydraulics. In this way she found occupations for herself that were separate from those of Skem, while expressing aspects of her own personality.

While she was busy on this, she did not neglect all her other creative work, not just bookbinding and woodengraving, but also dressmaking. Throughout her life, she made most of her own and her daughters' dresses, blouses, and even nightdresses from beautiful cotton lawn she bought at Liberty's in Regent Street, on an old black-and-gold Singer sewing machine, which she turned by hand. She could also do tailored things like jackets and skirts. She made William Morris chintz loose covers for the furniture and embroidered a striped seat cover for a particularly fine mahogany armchair. Apart from her enjoyment of these activities, she saw them as her way of contributing to making Skem's surroundings as comfortable and attractive as possible, which was another way in which she supported him in every aspect of life. He, in turn, quietly valued these skills of Nancy's, which were so very different from his own. When she asked him openly for his opinion, the most he tended to say was 'Very nice, Nancy, dear.' This was enough for her, and all she expected.

Chapter 8

The 80s and 90s – A Literary Retirement

Skem retired as head of department with some relief in 1976. Then, in 1981, aged 67, he retired as Professor of Civil Engineering. His decision to stay on for two years beyond normal retirement age was, in part, influenced by financial considerations. He had built up an unbroken record of pension contributions since 1936 and, as the amount of the pension was a proportion of the final salary, two extra years, in a time of wage inflation, were useful. The world of academia was also changing out of all recognition. The Thatcher era of cutting public expenditure had a huge impact on higher education. Gone are the days of the University Grants Committee distributing large sums for university expansion in the early 1960s. Staff now have to spend a great deal of time putting together bids for research funding and students struggle to survive on loans rather than grants. Skem can wax furious about the detrimental effect of these changes on academic independence and standards. At Buckingham Palace, on the day of his investiture, when the Queen commented that he must have seen many changes over his long career, he replied, 'Yes, and not all for the better, Ma'am.'

There was never any question of his leaving Imperial College. There was discussion with the Rector, Lord Flowers, and he stayed on in the post of Senior Research Fellow, as his predecessors had done. Pippard had been around for two or three years after his retirement. Of the generation after Skem, John Hutchinson and Peter Vaughan are now Senior Research Fellows and come into college weekly or fortnightly. Skem is still there 20 years later, in his own room. He continued with his postgraduate soil-mechanics lectures after his retirement and only stopped in 1995.

The artist chosen to paint Skem's formal portrait for the college (see frontispiece) in 1981 was Richard Foster. He was born in 1950 and tends to paint well-connected individuals sitting on striped sofas or in sunlit gardens.

The leaving party was organized by John Burland and Marjorie Carter, and held upstairs in the by now no longer new civil engineering building. Bernard Neal, Skem's successor as head of department, had asked Nancy what he would like as a gift and she chose an aquatint of a cutting on the Liverpool and Manchester railway in 1830, which now hangs in the hall at The Boltons. This was presented to him by Professor Neal at the party.

In the year of Skem's retirement, he gave his large collection of important early books and papers on soil mechanics and foundation engineering to the Civil Engineering Department library. Marjorie Carter and he drew up a bibliographical catalogue of the collection. He remembers finishing work on it while on holiday with Katherine at Gringley-on-the-Hill. Many of the books had been beautifully bound by Nancy. Photos of *Frontispieces* and sample pages of key works were taken by Joyce Gurr, the departmental photographer (a post which no longer exists in this age of digital cameras), and Skem was overjoyed that the book was so beautifully printed by John Cooper of the firm Dalbeattie of Croydon. The frontispiece is a colour photo of a selection of books from the collection, showing some of Nancy's bindings. The collection now forms a separate but integral part of the History Collection, joining the material Skem salvaged from the Civils in the 60s.

In the year of his retirement Skem was proud to be awarded the Karl Terzaghi Award, an honour as prestigious as the invitation to deliver the Rankine lecture. He was also awarded the Gold Medal of the Institute of Structural Engineers at a ceremony at the Hilton Hotel. The citation reads, 'for his contribution to the theory and practice of soil mechanics, particularly as they affect the foundations of buildings, and for his contributions to the history of structural engineering.' In his response Skem quoted, as he had done in his Inaugural Lecture, from William's Blake's poem. 'Auguries of Innocence' 'To see the world in a grain of sand,' adding, 'Or in my case, a particle of clay!' Julia, who attended the presentation comments, 'What a marvellous quote for a soil mechanic. Where did he get it from? He once told me that he and Nancy never read anything except detective novels. Most engineers are incredibly un-literate. It was at that moment I thought, how can I be so fortunate as to have a man like this in my life?' (I checked with the Institute but there is no record of this speech.)

Later that same year he was awarded an honorary doctorate of Chalmers University, Göteborg, Sweden to add to the doctorates which he'd been awarded at Durham University in 1968, and Aston in 1980.

During 1981, Skem and Nancy made several trips to see Sonia Rolt at Stanley Pontlarge, sometimes with Julia Elton and sometimes alone. On All Fools' Day, April 1st, they went out to Sharpness, at the end of the Gloucester and Berkeley Canal, where it joins the Severn estuary at a Telford dock. This is still navigated by canal boats and there is a canal museum. Later that summer they all went to a sort of 'coracle-fest' on the river Avon. They played about in coracles and took turns at paddling them. On later trips to Stanley Pontlarge, Skem remembers summer picnics and concerts in the tithe barn at Stanway, a beautiful house owned by the Earl of Weimess, where Julia played her oboe.

Smeaton book

The next literary undertaking for Skem was the editing of his book *John Smeaton, FRS*, published in 1981 by Thomas Telford Ltd. It was the first full account of his life and works since Samuel Smiles' *Lives of the Engineers* of 1861. Skem drew together chapters on different aspects of Smeaton's work, for example, Rowland Mainstone on the Eddystone Lighthouse, Charles Hadfield on his river navigations and canals, and Denis Smith on his mills, while he himself wrote the introduction and chapters on fen drainage and harbours. Taken together, they give a comprehensive understanding of Smeaton's achievements and place in the development of the profession of civil engineering. Many of the aspects of Smeaton highlighted by Skem in his introduction could apply as well to himself. 'Often his reports – though directed towards a very particular object, such as improving a harbour entrance – start with a discussion of the general principles involved and it is from these that he derives the practical solution.' 'He was as much at home on a construction site, working on the drawing board, or discussing a job with his clients or his resident engineer, as in making scientific experiments or reading a paper to the Royal Society.' The principle on which he says Smeaton based his practice is also his own. 'Civil engineering has to be both an art and a science, and the engineer's responsibility is to develop both aspects to the limit of his powers in order to fulfil the clients' requirements as safely and economically as possible.' He quotes from a letter from Smeaton of 1754, at the outset of his engineering career, saying that his principle is '… never to meddle in any Business, whether great or small, without Endeavouring to bring it to the utmost perfection of my Power'.

Smeatonian Society

Skem had long been a member of the Smeatonian Society and, in 1981, he was gratified by his appointment as president, securing him a place in a chain of civil engineers stretching back to the Society's foundation by John Smeaton and his circle. (Brown) It was set up in the 18th century as a gentlemen's dining club. At the first meeting, in 1771, the following resolution was passed: 'Agreed that the Civil Engineers of the Kingdom do form themselves into a Society consisting of a President, a Vice-President, Treasurer, and Secretary, and other Members who shall meet once a fortnight… at seven o'clock from Christmas or so soon as any of the country Members come to Town… to the end of the sitting of the Parliament.' The 18th-century atmosphere is preserved in the six dinners and one lunch that are held annually. Twenty to thirty senior civil engineers attend and Skem relishes the memorial toast that is drunk on each occasion, 'To the memory of our late worthy brothers, Mr Smeaton, Mr Mylne, Watt, and Mr Rennie.' The omission of 'Mr' seems to imply that Watt was not considered by his contemporaries to be a gentleman. A second toast is made to 'Waterways private and public, that contribute to the use, to the comfort, or the happiness of mankind.'

To continue the Smeaton theme, Skem worked with Mike Chrimes, librarian, and the Archives Panel of the Institution of Civil Engineers to mount an exhibition there in 1992 of drawings, plans, and photographs of Smeaton's works. The Institution's Archives Panel had been set up in 1975. A then vice-president, Sir Alfred Pugsley, was appalled at the way the Institution kept its own records and decided a panel should be set up to advise on the whole historical record of the Institution. Skem was the chair of this panel from its inception and for about 20 years.

It was also around that time that the Institution, in a misguided attempt to create wall space, was dismantling a comprehensive collection of portraits and paintings that had taken 150 years to build up. Mike Chrimes told me they were proposing to sell off the portraits for about £50 each. After their very first meeting, the Archives Panel sent a stiff note to the effect that these sales should stop immediately. 'Undoubtedly, a quite polite minute, but firm,' says Chrimes. 'As a chair, Skem can be very firm if he's not happy with the way things are going.' In Mike's view, 'He has clear views about what the Institution's role should be. For example, he has never been a campaigner for conservation of things. Giving advice is one thing, campaigning another.'

Bramianos dam

Skem visited the site of the Bramianos dam in Crete in the summer of 1980. This dam was to have a wide compacted silt core, and there was a risk of a flow of water under the silt causing erosion. A snapshot of an informal meeting under the trees shows Skem wearing muddy shoes and a very uncharacteristic baseball cap.

Salisbury Cathedral

Salisbury Cathedral was built in the late 13th century but the tower and the famous spire, which soars to a commanding height of 400 feet, were not added until a generation later. The spire is the tallest mediaeval structure in the world and is built of thin limestone slabs supported by an interior timber framework. These slabs carry their own load onto the tower. The tower and spire together added an enormous extra load onto the four great piers at the crossing of the aisle and transepts. The soil on which the cathedral sits is a layer of gravel over the underlying chalk of the Salisbury Plain. The foundations are only four feet deep. According to elementary theory available at the time, the whole structure should have failed when the tower was built. The dean and chapter of the cathedral had recently charged the firm Gifford & Sons of Southampton with stabilizing the tower. In 1982, Gifford called Skem in to advise as to how safe the tower foundations were. Skem worked with Gifford's senior engineer, Peter Taylor, then in early middle age, who was passionately devoted to the cathedral. (It is also one of Skem's favourites, and the place where he used to take foreign visitors, such as Ralph Peck.) Skem remembers driving with Peter each morning from his office in Salisbury towards the cathedral and, as the spire hove into view, Peter would sigh with relief, 'It's still there, thank goodness.' Skem liked Peter very much and was upset when he died suddenly.

'Admirable' observations had been made in the 18th century of the settlement of the tower piers in relation to the arches. There had been a big settlement in the distant past but this had stabilized by that time. Skem took measurements and made comparisons to those of the 18th century and found that only very small movement had occurred since then. He also had the benefit of some very large-scale model tests, which had recently been carried out in Germany in a research institute near Berlin, using two square metre test plates. As at Bath, his recommendation was to do nothing, to let well alone. Skem, who often had little time for church people, liked the Dean of Salisbury of the time very much. The Dean's response to Skem's recommendation is not recorded. (The Dean is now dead, and there is no record of discussions in minutes of Chapter meetings.)

Kalabagh

Early in 1982, Binnie & Partners were engaged by the World Bank to design and prepare contract documents for the Kalabagh Dam, on the River Indus in Pakistan. This was to be a dam on the same huge scale as that at Mangla. Binnie's led an international joint venture, including two Pakistani engineering firms. Initially, Skem was asked to appraise four sites on some 12 miles of the river and made his first visit in April 1982. At the time, very little was known about the geology around the proposed foundations but it was found to be similar to that at Mangla. Skem's report to Binnie's, written in conjunction with Alan Little, makes practical and pertinent, yet clear and simple statements on the seismicity, shear zones, uncemented sandstones, and the maximum recommended height for water-retaining structures. Again, a lab was set up on site, staffed by Pakistani engineers, where excellent work was done.

His second site visit, in June 1983, was made soon after the discovery of shear zones in the clay-stone strata at the selected dam site. It resulted in detailed recommendations about the procedure necessary to establish strength parameters for each stratum. He visited the site again in April 1984 to examine the data obtained since the previous June and to review the bedrock strengths to be used in design for both static and dynamic conditions. In extending the work on the residual strength of Siwalik clays that he had begun at Mangla, he was particularly interested in the effect of very rapid rates of shearing, such as might occur in an earthquake, on the strengths of the clay stones. He investigated this aspect using ring-shear apparatus at Imperial College, and brought in Ambraseys to advise on the use of rapid shear-test results in earthquake designs at Kalabagh.

Skem's view is that the soils-engineering problems at Kalabagh were much less controversial than at Mangla because the fundamental principles were now accepted. However, Gordon Eldridge of Binnie's, who was a close observer of Skem's Kalabagh work, considers that his investigation of the effect of rapid rates of shearing on the strength of clay stones was a notable extension of knowledge. He very much admires the thorough, logical development of his recommendations on design parameters and processes for design. 'It was a revelation for me.'

Again there was a consulting board but this time Skem took little part in it. The board did not argue with the engineering findings and designs were prepared, taking the shear zones into account. However, the Kalabagh Dam has not yet been built, for political rather than engineering reasons, as a result of problems of population resettlement and conflicts of interests between Punjab, the powerful Northwest Frontier regions upstream, and the downstream state of Sindh. Instead a barrage was designed on the Indus, below Tarbela.

Meanwhile, at Mangla, there was a huge flood in 1992, which stressed the dam and surrounding works to the maximum. The main spillway was operated fully but stood up to the volume of water. Unfortunately, some people, lulled into complacency by the success of the dam, had built houses on low-lying land downstream of the dam, and some of them lost their lives in those 1992 floods.

As a footnote, as far ahead as 2000, Binnie's were working on raising the Mangla Dam by 40 feet. The dam had been designed with this possibility, and therefore the higher water level, in mind. Binnie's again consulted Skem, sending him reports and asking him to review the 1965 test results, in conjunction with George Hallowes, the geotechnical man at Kalabagh. There was, on this occasion, no question of Skem, by now aged 86, travelling out to Pakistan.

June 4th 1984 was Skem's 70th birthday. On June 6th he gave a celebratory lecture to the British Geotechnical Society at the Civils, entitled 'Residual Strength of Clays in Landslides, Folded Strata, and the Laboratory'. Much of the lecture arose out of his work at Kalabagh. (*Géotechnique* 35. No.1 3–18,1985) John Burland flew in specially from Hong Kong to introduce him. He said, 'Professor Skempton's devotion to the history of civil engineering has by no means displaced his interest in current research and practice in soil mechanics. He has been remarkably active and, as a relative newcomer to Imperial College, it has been a wonderful experience for me to witness at first hand the intensity, the single-mindedness and the enthusiasm with which he has always worked and still works.' Arthur Penman was another who travelled a long distance to come to the lecture, in his case from the International Congress in Tokyo (missing the opportunity of post-congress study tours to see Japanese dams). He says that there was a dinner after the lecture at the St Stephen's Club for Skem and Nancy. Other guests included the Burlands, Dick Chandler, and

Jane Walbancke, with whom Skem had worked on Kalabagh, and the rest of the British Geotechnical Society committee.

Carsington

More or less as Skem was speaking at the Civils, there occurred a catastrophic failure at the Carsington Dam in Derbyshire. This is an earth dam that was being built by the engineering firm, G.H. Hill & Sons, which was very well-established in the Manchester and Pennine area. That June, when it was one metre short of its full height of 25 metres, a huge slip occurred on the upstream side of the dam. Within a day or two of the failure, the Severn Trent Water Authority asked Skem to investigate. Skem brought in his colleague Peter Vaughan of Imperial College, who had worked on several dams previously. Babtie Shaw & Morton of Glasgow, the consulting engineers of the investigation, set up a site office. Skem led the large-scale investigation. The findings were that a soft clay layer had been left in position under the upstream side of the dam, on the grounds that it would fully consolidate under construction, which it did. What had been missed was that there were solifluction shears in this clay that considerably reduced its strength. Peter Rowe of Manchester University had been involved in giving advice about the consolidation. Further, the clay core had been built with an upstream extension, an unusual if not unique design, which later calculations showed contributed to a severe reduction in stability. That it was the design that was at fault there is no question, but whether Hill & Sons could be blamed for making the mistake is another matter, because there were forces at work that were not fully understood by anyone at that time. This mistake, unfortunately, caused the ruin of the much-respected firm.

Skem made endless visits to Derbyshire all through that summer and beyond. He stayed in a hotel in Dovedale, grabbed an early breakfast, donned huge Wellington boots, and sloshed around the site taking samples. His preliminary results were available in October 1984 after four months' work, and his full report, written with David Coates, senior partner at Babtie's, came out in March 1985. The 500-metre slip had gone right through the core, for half the total length of the dam. It was beyond repair. It was necessary to demolish the entire dam. Peter Vaughan and Babtie's, advised by Skem, produced a complete redesign. Babtie's rebuilt the dam with Vaughan as technical adviser.

In the winter, water is pumped out of the River Derwent and into the now very large Carsington Reservoir via a long tunnel and, in the summer, water is sent back into the river in order to regulate the water supply for Derby. (The Derwent is a tributary of the Trent, rising in the Pennines.) The work on Carsington occupied Skem right up until 1987, but the resulting *Géotechnique* paper, written with Peter Vaughan, did not come out until 1993 (only because of the pressure of other work).

Computerization of soil mechanics

The final part of the Carsington investigations was a brilliant theoretical analysis by David Potts, done on computer, which revealed the progressive nature of the failure. Skem understands *what* Potts did, but not *how* he did it. Starting with the observed properties of the materials, Potts was able to calculate the height at which failure would have inevitably occurred, with a nearly exact result. Nothing like this analysis had been done before and it ushered in a new era in soil mechanics, using computer modelling. This effectively excludes Skem, who is not the least bit computer-literate.

His contemporary, Ralph Peck (in an interview with Elmo DiBiagio and Kaare Flaate of NGI in 2000) discusses the changing emphasis in engineering education in a passage that I'm sure Skem would echo: 'In my early days of teaching, soil mechanics was new and teachers in that field were in demand and generally in favour. Today, the great advances are in computational skills and modelling of behaviour of engineering structures. These advances are real and essential to the development of the profession. As with most advances, however, there is a downside. Many teachers, and consequently their students, come to think of the models as reality. This is dangerous and the danger is not always appreciated by the teaching professions… In geotechnics, the use of instrumentation and the observational method help considerably in keeping the subject close to reality.'

Nick Ambraseys contrasts the 'hands-on' approach taken by engineers from the 19th century up until the 1980s, with the computer-based science that now dominates contemporary civil engineering. 'The Victorian engineers learned from their failures. They built an aqueduct and it failed – they learned not to do the same again.' Skem's generation of engineers 'observed nature, studied geology, formulated the problem, and tried to discover the principles.' Now the computer has made feasibility studies redundant. The computer modelling is used to convince the client or the civil servants that the work can be done, that everything possible is being done. The client does not allow a proper analysis in the field. Failure is not allowed, and so the responsibility on the engineer is very great. 'Out in the field in Greece, eating our sandwiches with a glass of wine, we could review the morning's work much better than can be done at the meetings in hotels in Mayfair which now take place.' When he and Skem were on the Mornos job, for example, if some question arose, they could ask the driver to turn off the road and they would take another look.

In the light of these developments leaving him behind, and also in the face of the financial cuts imposed by the Government on higher education in the 1980s, in a sense, Skem retired at the right time. It would have been difficult for him, and possibly painful, to have continued in the now very changed worlds of civil engineering and academia.

Selected papers

In honour of Skem's 70th birthday, John Burland and Dick Chandler from IC took the initiative to prepare a collection of Skem's papers on soil mechanics, *Selected Papers on Soil Mechanics*, published by Thomas Telford, the publishing arm of the Institution of Civil Engineers. In his preface, Burland writes, 'In making the selection of papers we aimed to show Skempton's breadth of interest and achievement in the general field of soil mechanics and to include papers that the practising engineer, research worker, and student will find of value to have to hand. For the early papers, the selection was influenced by a historical interest, in addition to intrinsic merit. As well as advancing the subject, they reflect and are, on occasion, constrained by what was known at the time. This is well illustrated by the development of effective stress ideas in the context of the stability of slopes.' Burland omitted two major contributions – the Rankine lecture, because it is so readily available, and the classic paper with Don Macdonald of 1956 on the allowable settlements of buildings, which is too long to include. What comes out clearly from the collection is the extent to which consulting work has shaped Skem's research, and his ability to identify the complex geological features that control soil behaviour and are, therefore, crucial to the design of engineering works.

The book includes a short biographical essay by Silas Glossop and a piece by Bob Gibson on 'Working with Skempton' (See Ch. 4), both of which have been useful to me in writing this book.

San Francisco conference

In 1985 came the 11th ISSMFE International Conference, in San Francisco. This was to be the last of the four yearly conferences Skem went to. He presented a paper on the history of soil properties and also a report on Carsington with David Coates. Ralph Peck gave a keynote speech on 'The Last Sixty Years,' which reviewed the main strands of advancing soil-mechanics knowledge over that period, including a summary of Skem's work on pore pressures, triaxial testing, and clay properties. He ended with an appeal to uphold the observational method and the use of instrumentation in soil mechanics. He said, 'Sophisticated calculation is too often substituted for painstaking subsurface investigation. The ease and fascination of carrying out calculations taking into account complex loadings, geometrics, and soil conditions leads many of us to believe that realistic results will somehow emerge even if vital subsurface characteristics are undetected, ignored, or oversimplified. ...Not only does this practice lead to erroneous conclusions in specific instances, but it breeds a distaste for the painstaking fieldwork that may be required to disclose and evaluate those subsurface features that will determine safety and performance.' I can almost see Skem smiling in agreement in the conference audience.

Skem enjoyed staying in a comfortable hotel in vibrant downtown San Francisco and also going back to visit old friends at Berkeley but felt that, because the International Conferences had by this time become so enormous, they were very much less memorable than the earlier ones. There were 2000 people at San Francisco. (The Society has now changed its name to the International Society for Soil Mechanics and Geotechnical Engineering – ISSMGE.)

British civil engineering 1640–1840

During his research for the biographies of Jessop and Smeaton, and for his other historical writings and papers, Skem had been frustrated by the lack of bibliographical information on the printed reports and plans of the early engineers. Back in 1979, he had decided to embark on some systematic work to gather together the references into an accessible source. The Royal Society gave him a generous grant and he set off on his researches to the British Library, the Bodleian, the National Library in Edinburgh, and county record offices at Cambridge, Beverley, and Lincoln. In the middle of this, he had to undergo a hernia operation. Julia Elton and Mike Chrimes were very helpful, of course. Julia got heavily involved in the bibliography. Skem initially gave the index to a professional indexer. 'He said, would I just run my eye over it to make sure there weren't any howlers and I started going through it. So I was going into Skem's office twice a week with my offerings, having gone through the index, which it turned out wasn't going to do at all. Finally there was a moment when he looked at me and said, "I know what you're going to tell me," and I said "Yes, I think we really should start from scratch." And he said, would I do it? It took a lot of time, but it was lovely.' The index itself is a masterpiece of detail.

The bibliography of printed reports, plans, and books was published in 1987. (It is a much larger and more comprehensive bibliography than the small catalogue of his own personal collection, which he had given to IC on his retirement.) In his preface, Skem writes,

> It emerged that these reports, often finely printed and accompanied by engraved or lithographed plans, were far more numerous than could have been supposed and constituted a principal form of publication before the 1840s – before, that is to say, the time when they were, to an increasing extent, replaced in importance by papers published in engineering journals such as the *Transactions* of the Institution of Civil Engineers, first issued in 1836.

The book is now known in the trade simply as 'Skempton.' (Fifteen years on, Julia Elton and Roland Paxton are now in the process of compiling another bibliography of material 'not in Skempton'.)

In 1988, the British Geotechnical Association, publishers of *Géotechnique*, inaugurated the Skempton Gold Medal, awarded to a member who has made an outstanding contribution to the practice of geotechnical engineering over a sustained period. The aim is to honour those who have mainly worked in the United Kingdom, and there are no more than four awarded per decade. Penman was the recipient in 1997.

Nancy's decline

In the late 80s, to everyone's great distress, Nancy began to lose her short-term memory. Her wood engraving became wavery, one by one the saucepans in the kitchen became blackened as she burned food, and desperate lists in her now shaking handwriting appeared all over the flat as she tried to remember what she needed to get from the shops. The beds were unmade and a layer of dust filmed the antique furniture. I became aware that she was losing weight. This was distressing because she had had treatment for a malignant breast lump some time earlier. Skem always ate a nutritious lunch at college. He used to get some pâté and toast and grapes together for their evening meals, which was fine for him because he had had lunch. He assumed that Nancy fed herself during the day whereas, in fact, she was no longer able to do so. Skem is incapable of boiling the proverbial egg. For a time I tried to help by providing 'boil-in-the-bag' or readymeals for him to heat up in the oven but this was beyond him and it was all to no avail.

Skem treated Nancy with unwavering courtesy and respect and tried to continue to include her in social events as much as possible, causing some embarrassment to friends.

The normally enjoyable Newcomen field trips became an ordeal. Julia Elton remembers, 'When we were all in Sheffield for the Newcomen, he was staying with Nancy at Katherine's house and coming on field trips, and was meant to be going to a lecture and to the annual dinner. Sonia and I were supposed to be going round to spend the evening with them at Katherine's and he wrote us a note saying Nancy had taken a turn for the worse and he didn't think we ought to come. Katherine was in the loo with the phone clamped to one ear and Sonia and me in a public phone box in Sheffield. It was at that point Katherine said, "Can you try and keep his outside life as normal as possible?" And we certainly tried. Most of it was fetching and carrying. He simply didn't have the energy or inclination to do anything remotely difficult but, if you arrived on the doorstep to take him somewhere, then he would come.'

Things came to a head when Nancy collapsed rather dramatically at a dinner party at the house of Julia Elton and Frank Newby and had to be rushed by ambulance into Charing Cross Hospital. Skem was there until one or two o'clock in the morning. Julia and Frank were anxiously phoning him there but he kept insisting he could get a taxi home. Julia goes on, 'I'd never really talked back to him, but I said, "Don't be so silly, just ring, and we'll take you home. We're washing up, it's no problem." So we collected him, took him back to The Boltons, and poured ourselves large drinks. It was then that he said he wanted to get Nancy's book done. That very night was the first time he'd mentioned it. I guess it had always been in his mind and this experience had crystallized the idea. And he started on it right away.'

This book was a collection of the best of Nancy's wood engravings, which he arranged to have privately printed from the original blocks onto beautiful hand-made paper by the specialist Libanus Press in Marlborough. It was Julia Elton who suggested this publisher and she supported Skem by driving him down to Marlborough for meetings with Michael Mitchell of Libanus.

Julia gives a vivid description of these drives. Anyone who knows Skem knows that he has absolutely no small talk. Silences occur. Julia says, 'The silences have to be endured. I would pick him up, we'd interview the publisher, have lunch in some pub and, by the time I had done this three times there was nothing more to be said. Skem would sit drumming his fingers on the table. It always makes me twitch when he drums like that. I finally thought, I'm giving up an entire day of my life, I've found the right people to do the book, I'm doing the driving, do I also have to think of every single conversational gambit? So I did a little experiment with myself. We were driving back and he started to drum and I thought, I'm not going to start, I'm not going to twitch nervously, he can jolly well... Silence... After a bit he felt he had to say something and he started to tell me about you and Katherine. This had never happened before. I was fascinated, he had never talked about any of you. Suddenly I was hearing about his grandchildren. I thought, this is great!'

By the time the champagne was opened to celebrate the publication of the wood-engraving book, Nancy was only dimly aware of what was afoot but it was a lovely thing for him to have done. Nancy had willingly devoted her life to the support of Skem, deferring to his wishes, giving up art teaching because he asked her to, fitting both her mothering and her own work around the primacy of his demands, adapting to city life because that is where he needed to be. I take this book to be a token of his thanks to her for this and a most moving tribute to her once-rich, but now faded creativity.

Nancy was diagnosed with 'cerebral atrophy' and over the coming years she declined into a full-scale and devastating dementia. Skem sought domestic help through an agency and was introduced to a warm and capable Maori girl, Carolyn, who moved into the flat to care for Nancy. Through her, Skem met Bev Beattie, widow of a New Zealand farmer and ex-owner of a catering business, who had come over to England with her two daughters, Catherine and Louise. When one New Zealand family member went off on her travels, another moved in to help Skem.

Julia was a real source of support for Skem during that dreadful time with Nancy. He probably allowed this because she and Frank had seen Nancy at rock bottom. This support was undertaken at some considerable cost to herself, as this account shows: 'There were awful moments. Frank, at that period, was still seeing his wife once a month and I would ring up and say, "I'm all by myself. Can I come round and have supper?" This was a real ordeal because it was not at all easy with Nancy by this time. There was one dreadful night I went round to The Boltons and I did all the cooking on that terrible old gas cooker. He liked that. He got bored with whatever it was the carer did. I used to go to Marks and Spencer and buy all the things I could think of that he would like. Throughout one entire evening, Nancy said, every two minutes, "You do still love me, don't you?" Every time he said, "Yes, of course, I do, Nancy dear," and he did this the entire five hours I was there and I was filled with admiration. He was terribly sweet. You would never think – he has to be the least practical person ever and he had depended on her totally, so far as I could see. It was very, very touching and, actually, it was just about the only thing that kept me going because it was hell doing it. He's done so much for me, it was the least I could do.'

Skem and Nancy's golden wedding anniversary was marked by a family party at Hurlingham Club, but it was doubtful whether Nancy understood what was going on.

John Burland thought that the way Skem dealt with Nancy's illness was impressive. He says, 'He was always calm, never an angry word, very patient. He gained a great deal of respect at that time and people felt it indicated what a great man he was.'

For several years, the New Zealand network led by Bev Beattie took turns in the caring responsibilities until, eventually, Nancy had to move into a nursing home. Skem chose Meadbank, just over Battersea Bridge, so that he could visit regularly, and we set up Nancy's room with some of her paintings, in a vain attempt to remind ourselves, and her, of who she was. Julia intervened to get Skem out of the house and thinking about something less miserable than Nancy's illness. Robert Thorne was organizing a Victorian Society Conference at the Civils. The Victorian Society had previously concentrated on buildings and it was deemed valuable for the society to know about the great Victorian engineers, so Thorne ran a six-or-seven-week series of lectures in early 1992. Sonia and Julia rang Skem and said, 'Now, listen, come with us to the conference.' He havered and dithered but then said, 'Yes, I think I would rather like to do that.' They collected him and it was brilliant. He found the lecture enormously interesting and asked them back to The Boltons for a whisky.

The Thames Tunnel

During the years of Nancy's illness, there was a hiatus in Skem's work. He undertook no major engineering consultancy during that period, but Mike Chrimes of the Civils library and Julia Elton describe the work they, together with Skem, did on Marc Brunel's Thames Tunnel at Rotherhithe and how this came as a welcome diversion, taking his mind off the painful situation in his private life.

1993 was the 150th anniversary of the opening of the tunnel. Mike was hunting around for interesting things for the exhibition that the Civils was mounting, 'The Triumphant Bore'. Skem was supposed to write a major contribution to the catalogue but fell and cracked his pelvis and so was unable to do this, and Mike did the writing.

At about the same time, London Underground was concerned for the security and the safety of the tunnel (through which the East London Line runs). There were plans to cover the original brickwork with concrete. Skem had always been interested in the tunnel in terms of its scale and engineering difficulty. James Sutherland was called in and made new boreholes and investigations were carried out, so Skem had access to a quantity of geotechnical data and busied himself making a thorough study of the geology of the tunnel. The work Mike did for the exhibition came together with this technical material. Because 1994 was Skem's 80th birthday, it was decided the Thames Tunnel would be the subject of his birthday lecture at the Institution of Civil Engineers. He first gave an informal talk for the history group of the Institute of Structural Engineering, at which Julia reports he got very emotional about Marc Brunel and his famous tunnelling shield. 'He was passionate about it.' This was a 'dry run' for the birthday lecture, which was more formal. The lecture theatre at the Civils was packed, with a video link to surrounding rooms. Even as a non-engineer, I was riveted by his account of the geology through which Brunel had to tunnel and the hazards he encountered. Slides of his own cross-sections and of contemporary illustrations and portraits were clear and he had a mastery of modern audiovisual techniques to clarify his points.

The History Study Group of the Institution of Structural Engineers

Dr W.G.N. Geddes, who was president of the Institution of Structural Engineers in 1971–72, came from Glasgow and was worried about the extent of destruction of early industrial buildings there and elsewhere. He suggested that the Institution should be recording them.

James Sutherland was the original group convener of what became the Special Study Group on the History of Structural Engineering. The initial intention was to focus on the archaeology of structure but the group soon realized that they were more interested in the development of structural thinking, how things got designed, rather than simple archaeology, 'the written word rather than old remains'. James Sutherland says that, in the early days of the group, no-one else, except for the Newcomen Society, was discussing structural and civil engineering thought. Even the Newcomen was not doing so to any great extent, because it tended to concentrate, and still does, very much on mechanical and process engineering rather than on civil and structural engineering.

After 20 years, Sutherland decided it was time to hand over the group to someone else and, in 1993, Frank Newby took over as organizer. One lecture in the year is the formal 'Sutherland History Lecture'. The other meetings are

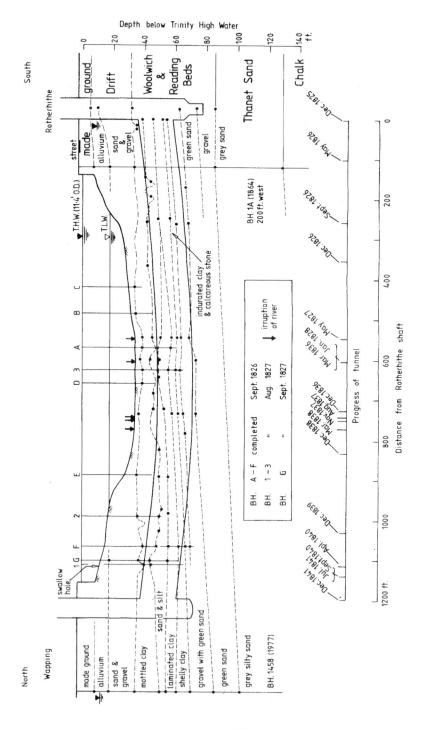

Geological section of the Thames Tunnel, from Géotechnique.

informal, where papers are often given as 'dry runs' prior to publication (as Skem's Thames Tunnel one was) and the speaker challenged by question and discussion from the floor. The group meets six or seven times a year and became very important to Skem. As late as February 2000, Skem gave a lecture on Samuel Wyatt to the History Group. 'As a structural engineer, he designed and built Albion Mill, Southwark (1783–86) with a raft foundation, prefabricated hospitals (1788), plans for completely iron-framed multi-storey buildings (1800), and the recently-discovered 90-foot-span iron-arch bridge at Culford Hall, Suffolk (1804), as well as several lighthouses.' He compared Wyatt with his contemporary, Robert Mylne, architect and civil engineer, whose books Julia had taken to Skem that first evening.

After the meetings, James Sutherland, Sonia Rolt, and Julia and Frank went back to The Boltons for a whisky. Julia comments, 'We arrive at the gate, and I say to Skem, "You are going to ask us in aren't you?" and Skem always says yes. It's actually very nice because they chew over engineering, and it seems to be his only social life, outside college.'

Sonia often travelled up from Gloucestershire and stayed with Skem at The Boltons on the night of the meeting. James Sutherland thinks that, since Nancy's illness, Skem became warmer and more human, his silences less formidable, and he was interested in a wider range of subjects. For example, in one evening discussion after the History Group, he expressed strong views about the parents in the case of Louise Woodward (the nanny accused of the murder of a child in her care in the USA), which James would not have expected some years ago. Skem is one of James' favourite friends.

An interest that Skem kept up during the years of Nancy's illness was Quaternary geology. He had been awarded the Geological Society Lyell Medal in 1972 in honour of his work in this field. He enjoyed the Quaternary Research Group's annual trips. Local members prepared the sites, cleaning up exposed strata, and so on. The trips lasted four or five days with evening talks and he made good friends among the geologists, especially Professor Fred Shotton.

He also made trips with the Geology Society's Engineering Group. He went to Sheffield one year, no doubt visiting Katherine. Then came Wales and Edinburgh. He remembers the biting cold in north Norfolk and another year it was Cornwall, where, after a day of constant blustery rain, during a walk along the cliffs, the rain began to blow almost vertically upwards. It penetrated Skem's waterproofs to such an extent that, when he eventually regained his hotel room, he could wring water out of his vest. The last trip he went on, in 1995, was to revisit Tilbury Marshes. Far from the oil refineries, the marsh has a fenlike quality that very much appeals to Skem. He wrote a paper, 'West Tilbury Marsh', for the Quaternary Research Association, Durham, that same year. He found that the trips gradually became populated by keen young men with reputations to make and they lost their appeal.

Silas Glossop.

Deaths

Nancy died in June 1993, on her 80th birthday, unable to recognize any of us, but Skem had been gradually losing the real Nancy for years before. He wept. He wanted no-one to come to the funeral except family, and no music, so we had rather a grim little service at Putney Crematorium, but he did want everyone to send flowers. He kept a meticulous note of who had sent wreaths, which he put in the back of his copy of the wood-engraving book.

That same year, Skem was also very saddened by the death of Silas Glossop. Since Silas' retirement and permanent move to the village of Brane, in the far west of his beloved Cornwall, Skem had seen much less of him and Sheila than he would have liked. He and Nancy had paid a few visits over the years and, one summer, they were briefly tempted by the prospect of buying a cottage near St Just. Silas had communicated to Skem some of his passion for the Neolithic sites and abandoned tin mines of the Cornwall inland landscape but, on mature consideration of the isolation, the long journey from London and the treeless and windswept countryside, they had wisely thought better of this idea. The gentle hills and beechwoods of the Chilterns are more to Skem's taste. He wrote a tribute to Glossop in *Géotechnique*. He was very pleased that the Glossop lecture has been set up by the Geological Society in Silas' memory. It is given annually at the Royal Institution lecture theatre and the speaker is presented with a medal by Silas' widow, Sheila, accompanied by his daughters,

Skem and Bev. Photo by Katherine Sisum.

Sue Chitty and Emma Slack. Peter Fookes, Skem's friend from Mangla days, gave the first lecture in 1997 and Dick Chandler the third. John Hutchinson gave the fourth in 2000, entitled 'Reading the Ground', on geomorphology and site appraisal.

Bev

After Nancy's death, it was arranged for Beverley Beattie to stay on at The Boltons. It worked very well as he was looked after and helped by her. She also looked after another nearby friend. During the years that followed we gradually became aware that Skem's relationship with Bev was changing.

He let it be known among close friends that they were now partners, and I know he would have liked to marry her. As Julia says, 'Bev was absolutely part of the scene. In fact when Sonia and I had our joint 80th and 50th birthday party at the Science Museum, Skem rang Sonia and said, could he bring Bev? So that was fine, she came as his partner. At first we couldn't quite figure out what to do, because they're together but not together. Nowadays I just say, would you like to bring Bev?' I don't know what prevented a marriage. One day he said to me, half jokingly, 'What would people think, an old codger like me marrying a so much younger woman?' This is not a point of view I would have expected to carry weight with him. Bev certainly brought great happiness and sense of security to his last years.

Ashgate Publishing

Towards the end of 1994, Dr John Smedley, the publisher of the Variorum collected papers series (an imprint of Ashgate Publishing), approached Joyce Brown, Skem's former secretary, and, by then, a senior lecturer at IC, to invite her to edit a twelve-volume series on the history of civil engineering. It comprises a set of collections of papers on the subject, republishing the papers in their original form. Joyce consulted Skem on whether he thought such a series was feasible. He was immediately enthusiastic and suggested the formation of an advisory panel, comprising himself, James Sutherland, Norman Smith, and Mike Chrimes, all engineering historians of different specialities. The panel discussed ways of dividing the subject into twelve, suggested the names of individual editors for each volume, and commented on finished volume proposals. 'It was a very good team,' says Joyce. Each volume editor's job was to choose about 16 or 18 already published papers and, in a scholarly introduction, set them in the context of the subject and its historiography.

Although volume editors had a free choice of papers for inclusion, papers by Skem appear in eight of the volumes, showing his major contributions to a number of historical topics: land drainage, clay embankment dams, dock engineering, river navigations, railway embankments and cuttings, iron-framed buildings, and early cements.

Joyce loved the literary aspect of it. As a child she enjoyed making books, creating anthologies, 'cutting out my favourite poems and binding them up, so making books appeals to me'.

As the work somewhat lengthily proceeded (it was a huge task), she and Skem had many discussions and she was grateful for his interest and support. As she says, it was useful to have someone look at the project objectively overall. The set of volumes is an important contribution to establishing the history of civil engineering as a subject in its own right and a major achievement by Joyce.

During this period, Dr Smedley asked Skem if he would consider contributing a volume to the Collected Papers series. The resulting book, *Civil Engineers and Engineering in Britain, 1600–1830*, published in 1996, contains 12 of his historical papers, updated in some cases by further research. The book could be seen as a historical parallel to the *Selected Papers on Soil Mechanics*. It brings together in one place material about an important phase in the evolution of the civil-engineering profession. In his introduction, Skem poses questions about pre-Victorian engineering history, which he goes on to answer in the papers. He also describes how the organization of civil-engineering works evolved over the period, with the emergence of the resident engineer, and arrangements for contracting and consultancy. James Sutherland likes the way that, alongside the engineering, Skem writes about the people and how they worked. Their characters emerge. For example, he transcribes the vivid correspondence between Wyatt and Boulton and Watt about the construction of the Albion Mill at Southwark. In an essay on the early members of the Smeatonian Society, he writes of Robert Mylne, who built the beautiful Blackfriars Bridge in London in 1769, that some of his later schemes were hampered by Mylne's 'irascible temperament. One of his workmen said that "Mr Mylne was a rare jintleman, but as hot as pepper and as proud as Lucifer".' Although he was also 'critical of his contemporaries', he was 'possessed of a talent amounting to genius as a structural and architectural designer'.

The idea for a *Biographical Dictionary of Civil Engineers* emerged out of the Archives Panel of the Institution of Civil Engineers. He and Mike Chrimes sat down one afternoon and drew up an initial list of about a hundred names, to which they subsequently added several more. Skem was chairman of the editorial board and worked on the project over recent years. Minor figures are given a brief paragraph, while the major players, such as Smeaton, have several hundred words. A large number of different contributors were asked to write entries on their specialist areas of study. Skem and Mike had certain difficulties in getting appropriate material from some of the contributors, to the extent of having to rewrite several of the entries.

In May 2000 we took Skem on a weekend break to Sutton on the Great Ouse, and he very much enjoyed revisiting Denver sluice and other fen drainage sites. At Ely Cathedral, to his great excitement, he discovered the handsome eighteenth century grave of Humphrey Smith, with its inscription, 'most Eminent in his Superior Abilities in Draining Fenny and Marsh Lands.' The work on the book is now complete, and a photo of this tomb is to grace Skem's entry on Smith in the Biographical Dictionary.

Knighthood

For some years, it had been felt in the civil engineering fraternity that Skem's achievements merited a knighthood. The usual process when someone is felt to be deserving of an honour is that a case is made to the appropriate government department, together with supporting testaments. (The prospective candidate is not usually informed, for fear of arousing expectations.) John Burland, David Coates, Mike Chrimes, and others had, for some time, been pressing through the 'system' for a knighthood to be awarded to Skem, without result. In successive years, Sir Ronald Oxburgh, the Rector of Imperial College, was urged to keep up the pressure, and Burland set about assembling written support from many quarters, completely without Skem's knowledge. A journalist got hold of Burland's note encouraging people to write, and Burland was not the only person to be devastated to read, in the edition of the magazine, *New Civil Engineer*, for March 18th 1999, a journalistic article implying that Skem not only knew of these efforts but endorsed them. Burland wrote a distressed letter to the editor, '… you did not give the slightest consideration to the feelings or reaction of this great and unassuming man for whom an honour is long overdue.' Skem's own reaction, as reported by Julia Elton, when she agitatedly phoned him, was phlegmatic. He hates all this kind of fuss. In fact, Julia is right when she says, 'A hallmark of Skem is his integrity. He is not on the make. He is rock-solid, pure, but uncompromising in his views.'

After this distressing setback, the cudgels were taken up again. John Burland received copies of letters of support from all over the soil-mechanics world that had been sent to the relevant private office. To quote from them may seem hagiographic but, taken together, they do sum up many of Skem's achievements. Sir William McAlpine wrote, 'The transformation in the design of the foundations of ground-bearing structures would not have occurred but for the skill and foresight of Professor A.W. Skempton. The science of soil mechanics was unknown much before the late 1930s but the demands of the planners and architects for high-rise buildings led to research into the properties and behaviour of soils under stress. Professor Skempton quickly became the UK's leading expert in this new discipline and, in 1945, he established a department at Imperial College, which was soon recognized as the leading establishment in the world, gaining Professor Skempton an international reputation which he still enjoys.

The Department… remains one of the most highly regarded for teaching and research in soil mechanics. The post-war development of many of the office and commercial buildings in London would not have been possible without the innovative foundation designs made possible by the meticulous and rigorous research undertaken by Professor Skempton'.

D.H. Cowie, Chairman of Binnie's (now Binnie Black & Veatch), writes that he 'helped to move geotechnical engineering from its infancy to a recognized discipline within the wider framework of civil engineering. His papers on foundations, stability of slopes, residual strength of clays, and fundamentals of soil behaviour are still definitive references throughout the profession'. Richard Jardine, Imperial College's present Professor of Geomechanics (as soil mechanics is now known) wrote, 'Skem is not only an engineer, scientist, and historian: he has been an accomplished sportsman, musician, and art enthusiast and remains interested in almost all areas of Western culture'.

These efforts bore eventual fruit and Skem was awarded a knighthood in the Millennium New Year's Honours list. He had always maintained that his FRS meant more to him than any other honour but, in the event, his 'K' gave him great pleasure and pride. Bev, Katherine, and I accompanied him to Buckingham Palace in June 2000, within a few days of his 86th birthday, and, despite his frailty, he comported himself with dignity before the Queen (unlike the comedian, Norman Wisdom, who received his knighthood immediately after him, 'accidentally-on-purpose' tripping up on the red carpet). The whole family then joined him for a celebratory lunch at the Athenaeum Club.

Earlier in April, John Burland and Mike Chrimes had organized an event

Collecting his knighthood from the Queen, 2000.

in honour of Skem's knighthood at the Institution of Civil Engineers. Friends, ex-students, colleagues, and family gathered in the grand surroundings of the Smeaton Room. In acknowledgement of Skem's love of music, Denis Smith put together a programme of musical items. Julia Elton played her oboe in a trio sonata by Telemann and then there was a selection of items on engineering-related themes, including a ridiculous Victorian parlour song about Brunel's Thames Tunnel, which had so involved Skem in the early 1990s. Everyone joined in the rousing chorus, 'Hail Great Brunel. The tunnel is the wonder of the world just now'.

At the turn of the millennium the *New Civil Engineer* pulled back some of the respect it had lost at Imperial College in the débâcle over the leak by nominating Skem as one of the greatest civil engineers of the 20th century, putting him, to Skem's scepticism, ahead of Terzaghi.

In November 2000 the Institution of Civil Engineers awarded Skem the Gold Medal for civil engineering excellence, recognising his 'contribution to the civil engineering profession over more than 60 years.' *New Civil Engineer* wrote, 'He has devoted more than 50 years to the development of Imperial College's centre of geotechnical engineering which has an international reputation… His work helped to make possible post-1950's construction of tall buildings in London clay.'

Ill health began to dog Skem in 1999. He tripped, while running for a bus, at South Kensington Station and fractured his pelvis. He made a full recovery from this, carrying out the prescribed physiotherapy exercises with characteristic determination and regaining full mobility very quickly. He had to undergo the unpleasantness of treatment for prostate trouble but overcame this also. Then came an affliction that was not so easily to be fought off – lung cancer. Like many of his generation, Skem was an enthusiastic pipe smoker all his life, resorting to cigarettes on occasion. He escaped the almost inevitable consequences until very late in life. He survived successive bouts of radiotherapy and returned to his desk and his almost daily walk to and from college. He bore the sessions in hospital, undergoing unpleasant tests and treatments with determination and, to use Sonia's word, 'gallantry'. Throughout these travails, Bev was an essential source of support, both practical and emotional.

As John Burland writes, 'Wherever one travels internationally, engineers ask after "Skem" and give their own personal testimonies about the influence he has had on their careers. There can be few engineers alive today who inspire such affection and who have made such a profound and lasting contribution in so many ways'.

Work undone

Some time after Skem wrote his study of the engineering works of John Grundy, (Originally published by the Society for Lincolnshire History and Archaeology in 1985 and reprinted in Civil Engineers & Engineering in Britain), the previously missing volumes of Grundy's engineering reports, drawings, and letters came to light at the University of Leeds. It represents a mass of new and valuable material. Mike Chrimes is sad that this find has come too late, for Skem was perhaps the only person who would be capable of making a comprehensive comparative evaluation of the information.

In the last weeks of his life he was at work, again with Mike Chrimes, on Robert Stephenson's London to Birmingham railway. This was the first really big trunk railway to be built in this country. Skem and Mike were doing a close analysis of the letting of the contracts, the way these contracts were managed, and the organisation of the construction of the railway in the mid-1830s.

For many years, Skem had talked of writing a definitive history of soil mechanics. He built up shelves of box files at The Boltons and at college. He has written substantial material on dams and started on cuttings and embankments. (His 1995 article on 'Embankments and cuttings on the early railways' for *Construction History* is considered by Dick Chandler to be 'the first authoritative article on this key aspect of railway construction.')

Mike Chrimes, who has helped him with this project, particularly at the time they were working on the Thames Tunnel (which could have been one of the case studies for the book), thinks that there are at least half a dozen unpublished papers awaiting completion and assembly into a book. Who will be able to pull this together?

Epilogue

Skem died in August 2001. The commentary to the *Selected Papers on Soil Mechanics* concludes with this description of 'the essential Skempton', which does, I think, identify aspects of my father's character that have been crucial to his success in his field. 'Single-mindedness to an unusual degree, great enthusiasm, and ability to inspire others, a natural appreciation of the important factors in complex situations, a proper sense of the indivisibility of nature, and a rare ability to work on until a clear synoptic view has been achieved'.

'Soil mechanics has received great benefit from the fortunate arrival of Skempton at a time when the subject was in a major phase of its development. It is also probably true to say that the geology of the country in which Skempton has worked has also been of considerable benefit to the subject. Just as the main progress in soft-clay studies has been made in the relatively homogeneous soft clays of Scandinavia and parts of North America, so it is not surprising that most of the concepts concerning the shear strength of stiff-fissured clays and their behaviour have been worked out (by Skempton) in studies of one of their more homogeneous and geologically uncomplicated members – the London Clay'. (Commentary to *Selected Papers on Soil Mechanics*)

In his 1985 address to the San Francisco conference, Ralph Peck summed up the progress made in soil mechanics in the quarter of a century since the first conference at Harvard in 1936 up until 1961, the period during which Skem was most active and innovative. He made a major contribution in all these areas:

- 'There has been a vast improvement in our understanding of the properties and behaviours of earth materials.
- Not only has soil mechanics been widely applied in practice, but field verification of behaviour has been actively pursued.
- Soil mechanics has been included in all civil-engineering curricula and its principles have become common practice.

- Numerous centres of activity in research and practice have sprung up around the world to meet local needs.
- Remarkable engineering works, probably not likely to have been possible without soil mechanics, had been accomplished'.

The historical side of Skem's researches is also summed up by Peck. 'He is recognized as one who instinctively cannot separate the design and construction of great engineering works from the personalities and thought processes of their creators or from the times in which the works were created; he can get into the minds of our predecessors and appreciate their accomplishments and he has shared this with his engineering colleagues in an exciting way'.

To add to these tributes to aspects of his work and professional achievements, I would add that, as a human being, his intelligence, his personal integrity, his honesty, his holding fast to all that he thinks right, his humour and his valuing of colleagues, friends, and his family, all mark him out as an exceptional man.

Mary, daughter of the great engineer John Smeaton, wrote on her father's death: 'The end he had through life deserved was granted; the body sunk but the mind shone to the last, and in the way good men aspire to, he closed a life, active as useful, amiable as revered'. The same can be said of my father.

Bibliography

References

(Other than Skempton's publications)

Allan, Harold. (Undated). *Fifty Years of Music at Imperial College*. Publ. Imperial College.
Beckett, H.E. (1968). *Bucknalls. A Short History*. Watford: Building Research Station.
Boughey, Joseph (1998). *Charles Hadfield. Canal Man and More*. Stroud: Sutton Publishing.
The Boltons Conservation Area Policy Statement (1981). Published by the Planning
    Committee of the Royal Borough of Kensington & Chelsea.
Brown, Joyce (ed.) (1985). *A Hundred Years of Civil Engineering at South Kensington*. London:
    Civil Engineering Dept. Imperial College.
Brown, Joyce (ed.) (1997). General Editor's Preface to *Studies in the History of Civil Engineers*.
    Aldershot: Ashgate Publishing.
Calder, Angus (1992 edit). *The People's War*. London: Pimlico.
Chitty, Susan. (1985). Now to my mother. *Weidenfeld & Nicholson*.
DiBiagio, E. and Flaate, K. (eds.) (2000). *Ralph B. Peck. Engineer, Educator, A Man of
    Judgement*. Oslo: Norwegian Geotechnical Institute. Publication No. 207.
Donnelly, Mark (1999). *Britain in the Second World War*. Routledge.
Farndale, Gen. Sir Martin (1986). *History of the Royal Regiment of Artillery. Western Front
    1914–18*. London: Royal Artillery Institution.
Glossop, R. (1977). *John Tunnard. A Personal Appreciation*. Catalogue of exhibition at the
    Royal Academy.
Goodman, Richard E. (1999). *Karl Terzaghi. The Engineer as Artist*. USA: ASCE Press.
Holroyd, Michael (1995). *Lytton Strachey*. London: Vintage.
Kolb, D.A. (1994). *Experiential Learning*. Prentice Hall.
Lee, Christopher (1999). *This Sceptred Isle: Twentieth Century*. London: BBC.
McCully, Patrick (1996). *Silenced Rivers. The Ecology and Politics of Large Dams*. London:
    Zed Books.
Microsoft (2000). *Snow, C.P., Baron Snow of Leicester*. http://encarta.msn.com.
The Newcomen Bulletin. **113**, April 1979 – Review.
Peat, A. and Whitton, B.A. (1997). *John Tunnard. His Life & Work*. Scholar Press.
Penman, A.D.M. (1998). *Geotechnical Engineering and Building Research. The Early Days of
    Soil Mechanics at BRS*. Unpublished paper.
Penman, Arthur. (2002). Professor Sir Alec Skempton and his Connection with Dams. In
    the journal *Dams and Reservoirs*.
Pevsner, N. (1960). *Pioneers of Modern Design*. Penguin Books.

Rolt, L.T.C. (1994 edition). *Narrow Boat*. Stroud: Sutton Publishing.
Skempton, A.W. (1989). *The Wood Engravings of Mary Skempton*. Marlborough: privately printed at Libanus Press.
Terzaghi. K and Frohlich, X. (1936). *Theorie der Setzung bau Taischichten*. Leipzig & Vienna: Deuticke.

Skempton's published work

Books

Terzaghi, K. (1960). *From Theory to Practice in Soil Mechanics* (eds. L. Bjerrum, A. Casagrande, R.B. Peck and A.W. Skempton). New York: John Wiley.
Institution of Civil Engineers (1969). *A Century of Soil Mechanics* (eds. L.F. Cooling, A.W. Skempton and A.L. Little). London: Institution of Civil Engineers.
Institution of Civil Engineers (1977). *Early Printed Reports and Maps (1665-1850) in the Library of the Institution of Civil Engineers*. London: Institution of Civil Engineers.
Hadfield, C. & Skempton, A.W. (1979). *William Jessop, Engineer*. Newton Abbot: David & Charles.
Skempton, A.W. (1981). *A Bibliographical Catalogue of the* [Skempton] *Collection of Works on Soil Mechanics 1764–1950*. London: Imperial College.
Skempton, A.W. (1981, reprinted 1991). *John Smeaton FRS*. London: Thomas Telford.
Skempton, A.W. (1984). *Selected Papers on Soil Mechanics*. London. Thomas Telford.
Skempton, A.W. (1987). *British Civil Engineering 1640–1840: a Bibliography of Contemporary Printed Reports, Plans and Books*. London: Mansell.
Skempton, A.W. (1996). *Civil Engineers and Engineering in Britain, 1600–1830*. Aldershot, Variorum.
Skempton, A.W. (2002). (editor and major contributor) *Biographical Dictionary of Civil Engineers, Vol.* **1**, 1600–1830. London: Thomas Telford Ltd.

Other published works by subject

Soil mechanics and geology

Skempton, A.W. (1938). Settlement analysis of engineering structures. *Engineering,* **146**, 403–406.
Cooling, L. F. & Skempton, A.W. (1941). Some experiments on the consolidation of clay. *J. Inst. Civ. Eng.* **16**, 381–398.
Cooling, L. F. & Skempton, A.W. (1942). A Laboratory Study of London Clay. *J. Inst. Civ. Eng.* **17**, 251–276.
Skempton, A.W. (1942). An investigation of the bearing capacity of a soft clay soil. *J. Inst. Civ. Eng.* **18**, 307–321.
Skempton, A.W. (1942). Some principles of foundation behaviour. *Journ. R.I.B.A.,* **50**, 3–6. (Also in *Building Science* (ed. D.D. Harrison), 41–51. London: Allen & Unwin, 1948).
Skempton, A.W. (1942). Soil Mechanics. In *A Geology for Engineers* by F.G.H. Blyth, 260–276. London: Arnold.
Skempton, A.W. (1942). Discussion: soil mechanics and site exploration, and soil mechanics in road and aerodromes construction. *J. Inst. Civ. Eng.,* **18**, 173–175.
Skempton, A.W. (1943). Discussion: Stress measurements in a cast-iron tunnel lining. *J. Inst. Civ. Eng.* **20**, 53–56.
Skempton, A.W. (1944). Notes on the compressibility of clays. *Q. J. Geol. Soc.,* **100**, 119–135.

A Particle of Clay

Skempton, A.W. (1944). Discussion: Stability analysis of Hollowell Dam. *Trans. Inst. Water. Engrs.*, **49**, 204–208.

Skempton, A.W. (1944). Discussion: Design of wharves on soft ground. *J. Inst. Civ. Eng.*, **22**, 34–37.

Skempton, A.W. (1945). A slip in the west bank of the Eau Brink Cut. *J. Inst. Civ. Eng.*, **24**, 267–287.

Skempton, A.W. and Glossop, R. (1945). Particle size in silts and sands. *J. Inst. Civ. Eng.*, **25**, 81–105.

Skempton, A.W. (1945). Earth pressure and the stability of slopes. In *Principles and Applications of Soil Mechanics*, 31–61. London: Institution Civil Engineers, 1946.

Skempton, A.W. (1948). A study of the geotechnical properties of some post-glacial clays. *Géotechnique*, **1**, 7–22.

Skempton, A.W. (1948). The $\phi = 0$ analysis of stability and its theoretical basis. *Proc. 2nd. Int. Conf. Soil Mech.*, **1**, 72–78. Rotterdam.

Skempton, A.W. (1948). The geotechnical properties of a deep stratum of post-glacial clay at Gosport. *Proc. 2nd. Int. Conf Soil Mech.*,**1**, 145–150. Rotterdam.

Skempton, A.W. & Golder, H.Q. (1948). The angle of shearing resistance in cohesive soils at constant water content. *Proc. 2nd. Int. Conf. Soil Mech.*, **1**, 185–192. Rotterdam.

Skempton, A.W. (1948). A study of the immediate triaxial test on cohesive soils. *Proc. 2nd. Int. Conf. Soil Mech.*, **1**, 192–196. Rotterdam.

Skempton, A.W. (1948). The rate of softening in stiff-fissured clays, with special reference to London Clay. *Proc. 2nd. Int. Conf. Soil Mech.*, **2**, 50–53. Rotterdam.

Skempton, A.W. and Golder, H.Q. (1948). Practical examples of the $\phi = 0$ analysis of stability of clays. *Proc. 2nd. Int. Conf. Soil Mech.*, **2**, 63–70. Rotterdam.

Skempton, A.W. (1948). A possible relationship between true cohesion and the mineralogy of clays. *Proc. 2nd. Int. Conf. Soil Mech.*, **7**, 45–46. Rotterdam.

Skempton, A.W. (1948). The effective stresses in saturated clays strained at constant volume. *Proc. 7th Int. Conf. App. Mech.*, **1**, 378–392. London.

Skempton, A.W. (1948). Vane tests in the alluvial plain of the River Forth near Grangemouth. *Géotechnique*, **1**, 111–124.

Skempton, A.W. (1949). Discussion: Site investigations. *J. Inst. Civ. Eng.*, **32**, 151–155.

Skempton, A.W. (1949). Discussion: Bearing capacity of screw cylinders in clay. *J. Inst. Civ. Engrs.* **34**, 76–81.

Skempton, A.W. and Bishop, A.W. (1950). The measurement of the shear strength of soils. *Géotechnique*, **2**, 90–108.

Skempton, A.W. (1952).The bearing capacity of clays. *Building Research Cong.*, **1**, 180–189. London.

Skempton, A.W. (1952). Discussion: Buoyant foundations in soft clay. *Proc. I.C.E. Part III*, **1**, 323–325.

Skempton, A.W. and Northey, R.D. (1952). The sensitivity of clays. *Géotechnique*, **3**, 30–53.

Skempton, A.W. and Ward, W.H. (1952). Investigations concerning a deep cofferdam in Thames estuary clay at Shellhaven. *Géotechnique*, **3**, 119–139.

Skempton, A.W. (1953). Soil mechanics in relation to geology. *Proc. Yorks. Geol. Soc.*, **29**, 33–62.

Gibson R. E., Skempton, A.W. and Yassin, A. A. (1953). Théorie de la force portante des pieu dans le sable. *Anns. Inst. Tech. Bâtim. Trav. Publics*, **16**, 285–290.

Skempton, A.W. (1953). The collodial 'activity' of clays. *Proc. 3rd Int. Conf Soil Mech.*, **1**, 57–61. Zurich.

Bibliography

Skempton, A.W. (1953). Discussion: Soil stability problems in road engineering. *Proc. Inst. Civ. Eng.*, Part II, **2**, 265–268.

Henkel, D.J. and Skempton, A.W. (1953). The post-glacial clays of the Thames estuary at Tilbury and Shellhaven. *Proc. 3rd Int. Conf. Soil Mech.*, **1**, 302–361, Zurich.

Skempton, A.W. (1954). Earth pressure, retaining walls, tunnels and strutted excavations. *Proc. 3rd Int. Conf. Soil Mech.*, **2**, 353–361, Zurich.

Skempton, A.W. (1954). Discussion: Settlement of pile groups in sand. *Proc. 3rd Int. Conf. Soil Mech.*, **3**, 172, Zurich.

Skempton, A.W. (1954). A foundation failure due to clay shrinkage caused by poplar trees. *Proc. Inst. Civ. Eng.*, Part 1, **3**, 66–86. Discussion Closure: 616–621.

Skempton, A.W. (1954). Discussion: Sensitivity of clays and the c/p ratio in normally consolidated clays. *Proc. Am. Soc. Civ. Engrs.* Separate 478, 19–22.

Skempton, A.W. and Bishop, A.W. (1954). Soils. In *Building Materials, their Elasticity and Plasticity* (ed. M. Reiner), 417–482. Amsterdam: North Holland.

Henkel, D.J. and Skempton, A.W. (1954). A landslide at Jackfield, Shropshire, in a heavily over-consolidated clay. *Proc. Conf. Stability of Earth Slopes*, **1**, 90–101. Stockholm; also *Géotechnique*, **5**, 1955, 131–137.

Skempton, A.W. (1954). The pore pressure coefficients A and B. *Géotechnique*, **4**, 143–147.

Skempton, A.W. & Bishop, A.W. (1955). The gain in stability due to pore-pressure dissipation in a soft clay foundation. *Trans. 5th Int. Cong. Large Dams*, **1**, 613–638. Paris.

Skempton, A.W., Peck, R.B. and Macdonald, D.H. (1955). Settlement analyses of six structures in Chicago and London. *Proc. Inst. Civ. Eng.*, Part 1, **4**, 525–544.

Skempton, A.W. (1955). Foundations for high buildings. *Proc. Inst. Civ. Eng. Part III*, **4**, 246–269. Discussion Closure: 305–308, 313.

Skempton, A.W. and Macdonald, D.H. (1955). A survey of comparisons between calculated and observed settlements of structures on clay. *Proc. Conf. Correlation of Calculated and Observed Stresses and Deformation in Structures.* 318–337. London. Discussion Closure: 783–784.

Skempton, A.W. (1955). *Soil mechanics and its place in the university*. Inaugural lecture as Professor of Soil Mechanics, Imperial College. London.

Skempton, A.W. and Macdonald, D.H. (1956). The allowable settlements of buildings. *Proc. Inst. Civ. Eng.*, Part III, **5**, 727–768.

Skempton, A.W. (1956). Discussion: Particle separation in clays. *Géotechnique*, **6**, 193–194.

Skempton, A.W. (1956). Discussion: Reconstruction of entrance lock, Royal Docks, Port of London. *Proc. Inst. Civ. Eng. Part II*, **5**, 164–165.

Skempton, A.W. and Henkel, D.J. (1957). Tests on London Clay from deep borings at Paddington, Victoria and the South Bank. *Proc. 4th Int. Conf. Soil Mech.*, **1**, 100–106, London.

Skempton, A.W. and Delory, F.A. (1957). Stability of natural slopes in London Clay. *Proc. 4th Int. Conf. Soil Mech.*, **2**, 378–381.

Skempton, A.W. and Bjerrum. L. (1957). A contribution to the settlement analysis of foundations on clay. *Géotechnique*, **7**, 168–178.

Skempton, A.W. (1957). Discussion: Consolidation of Usk Dam fill. *Proc. Inst. Civ. Eng.*, **7**, 267–269.

Skempton, A.W. (1957). Discussion: Further data on the c/p ratio in normally consolidated clays. *Proc. Inst. Civ. Eng.*, **7**, 305–307.

Skempton, A.W. (1958). Discussion: Design and performance of Sasumua Dam. *Proc. Inst. Civ. Eng.*, 9, 344–348.

Skempton, A.W. (1959). Discussion: Heave of London Clay: correlation with laboratory tests. *Géotechnique,* **9,** 145–146.

Skempton, A.W. (1959). Cast *in situ* bored piles in London Clay. *Géotechnique,* **9,** 153–173.

Skempton, A.W. (1959). Discussion: Soil engineering at Steep Rock Iron Mines, Ontario Canada. *Proc. Inst. Civ. Eng.* **13,** 93–95

Skempton, A.W. (1960). Effective stress in soils, concrete and rocks. In *Pore Pressure and Suction in Soils,* 4–16. London: Butterworths.

Skempton, A.W. and Henkel, D.J. (1960). Field observations on pore pressure in London Clay. In *Pore Pressure and Suction in Soils,* 81–84. London: Butterworths.

Skempton, A.W. (1960). Discussion: The pore pressure coefficient in saturated soils. *Géotechnique,* **10,** 186–187.

Skempton, A.W. (1960). Discussion: Cape Town Wemmershoek water scheme. *Proc. Inst. Civ. Eng.,* **15,** 149–150.

Skempton, A.W. (1961). Horizontal stresses in an over-consolidated Eocene clay. *Proc. 5th Int. Conf. Soil Mech.,* **1,** 351–357. Paris.

Skempton, A.W. (1961). Address of the President. *Proc. 5th Int. Conf. Soil Mech.,* **3,** 39–42.

Skempton, A.W. (1961). Discussion: Stability analysis of clay slopes. *Proc. 5th Int. Conf. Soil Mech.,* **3,** 349–350.

Skempton, A.W. and Brown, J.D. (1961). A landslide in boulder clay at Selset, Yorkshire. *Géotechnique,* **11,** 280–293.

Skempton, A.W. (1962). Discussion: An introduction to alluvial grouting. *Proc. Inst. Civ. Eng.,* **23,** 705–707.

Skempton, A.W. and Catin, P. (1963). A full-scale alluvial grouting test at the site of Mangla Dam. In *Grouts and Drilling Muds in Engineering Practice,* 131–135. London: Butterworths.

Skempton, A.W. and Sowa, V.A. (1963). The behaviour of saturated clays during sampling and testing. Géotechnique, **13,** 269–290.

Henkel, D.J., Knill, J.L., Lloyd, D.G. and Skempton, A.W. (1964). Stability of the foundations of Monar Dam. *Trans. 8th Cong. Large Dams,* **1,** 425–441. Edinburgh.

Skempton, A.W. (1964). Long-term stability of clay slopes. *Géotechnique,* **14,** 77–101. (Fourth Rankine Lecture).

Skempton, A.W. and La Rochelle, P. (1965). The Bradwell slip: a short-term failure in London Clay. *Géotechnique,* **15,** 221–242.

Skempton, A.W. (1965). Discussion: On the relevance of laboratory tests to slope stability problems. *Proc. 6th Int. Conf. Soil Mech.,* **3,** 278–280. Montreal.

Skempton, A.W. (1965). Discussion: Creep in clay slopes. *Proc. 6th Int. Conf. Soil Mech.,* **3,** 551–552. Montreal.

Skempton, A.W. (1966). Bedding-plane slip, residual strength and the Vaiont landslide. *Géotechnique,* **16,** 82–84.

Skempton, A.W. (1966). Some observations on tectonic shear zones. *Proc. 1st Cong. Int. Soc. Rock Mech.,* **1,** 329–335. Lisbon.

Skempton, A.W. (1966). Large bored piles: Summing up. *Symp. on Large Bored Piles,* 155–157. London: Inst. Civ. Eng.

Binnie, G.M., Clark, J.F.F. and Skempton, A.W. (1967). The effect of discontinuities in clay bedrock on the design of dams in the Mangla Project. *Trans. 9th Int. Cong. Large Dams,* **1,** 165–183. Istanbul.

Skempton, A.W. and Petley, D.J. (1967). The strength along structural discontinuities in stiff clays. *Proc. Géotech. Conf.,* **2,** 29–46. Oslo.

Bibliography

Skempton, A.W. (1968). Discussion: Geology of the Mangla Dam Project. *Proc. Inst. Civ. Eng.,* **1**, 133–137.

Skempton, A.W., Schuster, R.L. and Petley, D.J. (1969). Joints and fissures in the London Clay at Wraysbury and Edgware. *Géotechnique,* **19**, 205–217.

Skempton, A.W. and Hutchinson, J.N. (1969). Stability of natural slopes. *Proc. 7th Int. Conf. Soil Mech.,* **2**, 291–340. Mexico City.

Skempton, A.W. and Hutchinson, J.N. (1969). General Report on stability of slopes and embankment foundation. *Proc. 7th Int. Conf. Soil Mech.,* **3**, 151–155. Mexico City.

Skempton, A.W. (1970). The consolidation of clays by gravitational compaction. *Q. J. Geol. Soc.,* **125**, 373–411.

Skempton, A.W. (1970). First time slides in over-consolidated clays. *Géotechnique,* **20**, 320–324.

Skempton, A.W. and Petley, D.J. (1970). Ignition loss and other properties of peats and clays from Avonmouth, King's Lynn and Cranberry Moss. *Géotechnique,* **20**, 343–356.

Guidi, C.C., Croce, A., Polvani, G., Schultze, E. and Skempton, A.W. (1971). Caratteristiche geotecniche del sottosuolo della torre. In *Richerche e Studi su la Torre Pendente di Pisa,* **1**, 179–199. Florence.

Early, K.R. and Skempton, A.W. (1972). Investigations of the landslide at Walton's Wood, Staffordshire. *Q. J Eng. Geol.,* **5**, 19–42.

Chandler, R.J. and Skempton, A.W. (1974). The design of permanent cutting slopes in stiff fissured clays. *Géotechnique,* **24**, 457–466.

Skempton, A.W. (1976). Valley slopes: Introductory remarks. *Phil. Trans. Roy. Soc.* A, **283**, 423–426.

Skempton, A.W. and Weeks, A.G. (1976). The Quaternary history of the Lower Greensand escarpment and Weald Clay vale near Sevenoaks, Kent. *Phil. Trans. Roy. Soc.* A, **283**, 493–526.

Chandler, R. J., Kellaway, G.A., Skempton, A.W. and Wyatt, R.J. (1976). Valley slope sections in Jurassic strata near Bath, Somerset. *Phil. Trans. Roy. Soc.* A, **283**, 527–556.

Skempton, A.W. (1978). Stability of cuttings in brown London Clay. *Proc. 9th Int. Conf. Soil Mech.,* **3**, 261–270. Tokyo.

Skempton, A.W. (1985). Residual strength of clays in landslides, folded strata and the laboratory. *Géotechnique,* **35**, 3–18.

Skempton, A.W. and Coats, D.J. (1985). Carsington Dam failure. *Symp. on Failures in Earthworks,* 201–220. London, Inst. Civ. Eng.

Skempton, A.W. (1986). Standard penetration test procedures and the effects in sands of overburden pressure, relative density, particle size, ageing and overconsolidation. *Géotechnique,* **36**, 425–447.

Skempton, A.W., Leadbeater, R.D. and Chandler, R.J. (1986). The Mam Tor landslide, North Derbyshire. *Phil. Trans. Roy. Soc.* A, **329**, 503–547.

Skempton, A.W. (1988). Geotechnical aspects of the Carsington Dam failure. *Proc. 11th Int. Conf. Soil Mech. & Foundation Engineering,* **5**, 2581–2591. San Francisco.

Skempton, A.W., Norbury, D., Petley, D.J. and Spink, J.W. (1991). Solifluction shears at Carsington, Derbyshire. *Quaternary Engineering Geology* (Geol. Soc. Engineering Geology Special Publication No. 7), 381–387.

Skempton, A.W. and Vaughan, P.R. (1993). The failure of Carsington Dam. *Géotechnique,* **43**, 151–173.

Skempton, A.W. and Brogan, J.M. (1994). Experiments on piping in sandy gravels. *Géotechnique,* **44**, 449–460.

A Particle of Clay

Skempton, A.W. (1995). West Tilbury Marsh. *The Quaternary of the Lower Reaches of the Thames*. (ed. D.R. Bridgland, B. Allen & B.A. Haggart) Quaternary Research Association, Durham, 323–328.

Engineering history, biography and other subjects

Skempton, A.W. (1946). Alexandre Collin (1808–1890), pioneer in soil mechanics. *Trans. Newcomen Soc.,* 25 (1945–47), 91–104. Also, in translation by Professor J. Kerisel, *Annales des Ponts et Chaussés,* **126,** 317–340. (1956).

Skempton, A.W. (1949). Alexandre Collin: a note on his pioneer work in soil mechanics. *Géotechnique,* **1,** 216–221.

Skempton, A.W. (1952). Discussion: The historical development of structural theory. *Proc. Inst. Civ. Eng. Part III,* 1, 402–405.

Skempton, A.W. (1953). Discussion: Design of a reinforced-concrete factory. *Proc. Inst. Civ. Eng. Part III,* 2, 390–391.

Skempton, A.W. (1953). Engineers of the English river navigations, 1620–1760. *Trans. Newcomen Soc.,* **29** (1953–55), 25–54.

Skempton, A.W. (1956). Alexandre Collin and his pioneer work in soil mechanics. In *Landslides in Clays, by Alexandre Collin: 1846.* (trans. W.R. Schriever), xi–xxxiv. Toronto University Press.

Skempton, A.W. (1956). The origin of iron beams. *Actes 8eme Cong. Int. d'Hist. Sciences,* **3,** 1029–1039.

Skempton, A.W. and Johnson, H.R. (1956). William Strutt's cotton mills, 1793–1812. *Trans. Newcomen Soc.,* **30** (1955–57) 179–205.

Skempton, A.W. (1957). Canals and river navigations before 1750. In *History of Technology* (ed. C. Singer, *et al.)* **3,** 438–450. Oxford University Press.

Skempton, A.W. and Buckle, R. (1957). *Catalogue of exhibition of works of Thomas Telford, FRS, 1757–1834.* London: Inst. Civ. Eng., 32 pp.

Skempton, A.W. (1958). Arthur Langtry Bell (1874–1956) and his contribution to soil mechanics. *Géotechnique,* **8,** 143–157.

Skempton, A.W. (1959). The evolution of the steel-frame building. *The Guilds' Engineer,* **10,** 37–51.

Skempton, A.W. (1960). Terzaghi's discovery of effective stress. In *From Theory to Practice in Soil Mechanics.* (eds. L. Bjerrum, A. Casagrande, R.B. Peck & A.W. Skempton) 45–53. New York: Wiley.

Skempton, A.W. (1960). The Boat Store, Sheerness (1858–60) and its place in structural history. *Trans. Newcomen Soc.,* **32** (1959–60) 57–75.

Skempton, A.W. and de Maré, E. (1961). The Sheerness Boat Store 1858–60. *Journ. RIBA,* **68,** 318–324.

Skempton, A.W. and Johnson, H.R. (1962). The first iron frames. *Architectural Review,* **131,** 175–186.

Skempton, A.W. (1962). Portland Cements, 1843–1887. *Trans. Newcomen Soc.,* **35** (1962–63), 117–151.

Skempton, A.W. (1962). The instrumental sonatas of the Loeillets. *Music and Letters,* **43,** 206–217.

Skempton, A.W. (1970). Alfred John Sutton Pippard, 1891–1969. *Biogr. Mem. Fellows Roy. Soc.,* **16,** 463–478.

Skempton, A.W. (1971). *The Smeatonians: Duo-Centenary Notes on the Society of Civil Engineers 1771–1971.* London: Inst. Civ. Engineers

Skempton, A.W. (1971). The publication of Smeaton's Reports. *Notes & Records Roy. Soc.,* **26,** 135–155.

Skempton, A.W. (1971). The Albion Mill foundations. *Géotechnique,* **21,** 203–210.

Skempton, A.W. (1971). Samuel Wyatt and the Albion Mill. *Architectural History* **14,** 53–73.

Skempton, A.W. and Wright, E. (1971). Early members of the Smeatonian Society of Civil Engineers. *Trans. Newcomen Soc.,* **44** (1971–72), 23–42.

Skempton, A.W. and Brown, J. (1972). John and Edward Troughton, mathematical instrument makers. *Notes & Records Roy. Soc.,* **27,** 233–262.

Skempton, A.W. (1972). Laurits Bjerrum 1918–1973: a tribute. *Géotechnique,* **23,** 315.

Skempton, A.W. (1974). William Chapman (1749–1832), Civil engineer. *Trans. Newcomen Soc.,* **46,** (1973–74) 45–83.

Skempton, A.W. (1975). A history of the steam dredger, 1797–1830. *Trans. Newcomen Soc.,* **47,** (1974–76), 97–115.

Skempton, A.W. (1975). Notes on the origins and early years of the British Geotechnical Society. *Géotechnique,* **25,** 635–646.

Skempton, A.W. (1977). The engineers of Sunderland harbour, 1718–1817. *Industrial Archaeology Review,* **1,** 103–125.

Skempton, A.W. (1977). Dr. Leonard Frank Cooling, 1903–1977. *Géotechnique,* **27,** 265–270.

Skempton, A.W. and Andrews, A. (1977). Cast iron edge rails at Walker Colliery 1798. *Trans. Newcomen Soc.,* **48,** (1976–77), 119–122.

Skempton, A.W. (1979). Engineering in the Port of London, 1789–1808. *Trans. Newcomen Soc.,* **50** (1978–79), 87–108.

Skempton, A.W. (1979). Telford and the design for a new London Bridge. In *Thomas Telford, Engineer* (ed. A. Penfold) 62–83. London: Thomas Telford Ltd.

Skempton, A.W. (1981). Landmarks in early soil mechanics. *Proc. 7th European Conf. Soil Mech.,* **5,** 1–26. Brighton.

Skempton, A.W. (1981). Casagrande: one of the great civil engineers. *New Civ. Engr.* 1 Oct., 14.

Skempton, A.W. (1982). Engineering in the Port of London, 1808–1834. *Trans. Newcomen Soc.,* **53,** (1981–82), 73–94.

Skempton, A.W. (1984). The engineering works of John Grundy (1 719–1783). *Lincolnshire History & Archaeology,* **19,** 65–82.

Skempton, A.W. (1984). Engineering on the Thames navigation, 1770–1845. *Trans. Newcomen Soc.,* **55** (1983–84), 153–174.

Skempton, A.W. (1984). Engineering on the English river navigations in 1760. In *Canals: a New Look* (ed. Mark Baldwin & Anthony Burton) 23–44. Chichester: Phillimore.

Skempton, A.W. (1985). A history of soil properties. *Proc. 11th Conf. Soil Mech. & Foundation Engineering.* Golden Jubilee Volume, 95–121. San Francisco.

Skempton, A.W. (1989). Biographical Notes. In *The Wood Engravings of Mary Skempton.* Privately published, Libanus Press, Marlborough.

Skempton, A.W. (1990). Historical development of British embankment dams to 1960. *Proc. Conf. on Clay Barriers for Embankment Dams.* 15–52. London: Thomas Telford Ltd.

Skempton, A.W. (1993). Rudolph Glossop, 1902–1993. *Géotechnique,* **43,** 623–625.

Skempton, A.W. and Chrimes, M.M. (1994). Thames Tunnel: geology, site investigation and geotechnical problems. *Geotechnique,* **44,** 191–216.

Skempton, A.W. (1995). Embankments and cuttings on the early railways. *Construction History,* **11,** 33–49.

Index of names

Index of names